The Discovery of Time

The Discovery of Time

Stephen Toulmin
June Goodfield

THE UNIVERSITY OF CHICAGO PRESS
CHICAGO AND LONDON

For Leslie Farrer-Brown

The University of Chicago Press, Chicago 60637
The University of Chicago Press, Ltd., London

Originally published in 1965 by Harper & Row, Publishers, Inc.
Midway reprint 1977
Phoenix edition 1982
Printed in the United States of America

89 88 87 86 85 84 83 82 1 2 3 4 5

ISBN: 0-226-80842-4 LCN: 81-71398
Published by arrangement with Harper & Row, Publishers, Inc.

Contents

Plates

BETWEEN PAGES 144 AND 145

Acknowledgements

For permission to reproduce extracts from passages in copyright, the authors are grateful to the following:
Cambridge University Press for *De Rerum Natura* by Lucretius, translated by R. C. Trevelyan, and *John Ray, Naturalist* by Charles Raven; Jonathan Cape Ltd for *A Book of Beasts* by T. H. White; Columbia University Press for *The History of History* by J. T. Shotwell; Cornell University Press for *G. B. Vico, The New Science*, translated and edited by T. C. Bergin and M. H. Fisch; Faber & Faber Ltd for *Prospero's Cell* by Lawrence Durrell and *Philosophical Problems of Nuclear Science* by W. Heisenberg; the Johns Hopkins Press for *Essays in the History of Ideas* by A. O. Lovejoy and *Forerunners of Darwin* edited by B. Glass, O. Temkin and W. L. Strauss; the Loeb Classical Library for *Diodorus Siculus Vol. 1*; Macmillan & Co. Ltd for *The Idea of Progress* by J. B. Bury; Thomas Nelson & Sons Ltd for *Descartes: Philosophical Writings*, translated and edited by G. E. M. Anscombe and P. T. Geach; Penguin Books Ltd for the 'Canticle of All Created Things' by Francis of Assisi in the *Penguin Book of Italian Verse*; Routledge & Kegan Paul Ltd for *The Historical Revolution* by F. Smith Fussner; Mr Siegfried Sassoon for the sonnet 'Grandeur of Ghosts'; SCM Press Ltd for *Metaphysical Beliefs* by Stephen Toulmin; and Yale University Press for *Lorenzo Valla's Treatise on the Donation of Constantine*, translated and edited by Christopher B. Coleman.

When I have heard small talk about great men
I climb to bed; light my two candles; then
Consider what was said; and put aside
What Such-a-one remarked and Someone-else replied.

They have spoken lightly of my deathless friends
(Lamps for my gloom, hands guiding where I stumble),
Quoting, for shallow conversational ends,
What Shelley shrilled, what Blake once wildly muttered. . . .

How can they use such names and be not humble?
I have sat silent; angry at what they uttered.
The dead bequeathed them life; the dead have said
What these can only memorize and mumble.

Grandeur of Ghosts—SIEGFRIED SASSOON

Some write a narrative of wars, and feats
Of heroes little known; and call the rant
An history: . . . Some drill and bore
The solid earth, and from the strata there
Extract a register, by which we learn
That he who made it, and reveal'd its date
To Moses, was mistaken in its age.
Some, more acute, and more industrious still,
Contrive creation; travel nature up
To the sharp peak of her sublimest height,
And tell us whence the stars. . . . And thus they spend
The little wick of life's poor shallow lamp,
In playing tricks with nature, giving laws
To distant worlds, and trifling in their own.

William Cowper, *The Task, Book III* (1785)

By the present time, one can be certain that any theory which successfully explains the entire Past will inevitably—through that achievement alone—win the intellectual leadership of the Future.

Auguste Comte, *Discours sur l'Esprit Positif* (1844)

Authors' Foreword

With the appearance of this volume we have completed the first of the two tasks that we set ourselves in planning our analysis of *The Ancestry of Science*. This aim was never (as one reviewer caricatured it) to summarize 'the whole of the history of science' in four middle-sized books—an impossible task; but was rather to trace out the main lines of intellectual development by which the fundamental features of our contemporary picture of Nature came to be accepted as common knowledge. Having surveyed elsewhere men's changing ideas about the layout, workings and occupants of the natural world, we are here concerned with its 'life-story', and have attempted to depict the gradual emergence of a continuing sense of history out of earlier mythological and theological systems. In our final volume, we plan to set the intellectual genealogies examined in the earlier books into their own historical contexts, and consider what can be done to characterize in general terms the processes of 'give and take' by which scientific thought has continually interacted with the wider social and cultural environment.

In one respect this book differs from *The Fabric of the Heavens* and *The Architecture of Matter*. There, we were mapping the development of ideas whose place was securely within the natural sciences, and which overlapped only marginally into other regions of intellectual enquiry. In the area of historical ideas, by contrast, we have found ourselves continually driven outwards from natural science proper into other disciplines. Throughout the centuries of intellectual endeavour, the growth of men's historical consciousness across subjects ranging from physical cosmology at one extreme to theology and social history at the other, took closely parallel forms. Just *how* closely parallel, is the first of the two discoveries we hope to have demonstrated here. The other has to do with the recurrent patterns of theory manifested in all the historical sciences—whether those of Man or of Nature: where disciplines with quite different subject-matters have faced common forms of problem (e.g. the problem of establishing a well-founded temporal sequence of past epochs), they have—it seems—resorted again and again to similar strategies. To this extent, we conclude, physical cosmologists today may have more to learn than they yet recognize from the theoretical quandaries facing their predecessors in geology, and even in political theory.

The debts we owe to those colleagues from whose scholarly achievements we have profited in reconstructing this 'history of history' accordingly range more widely than before. Much of the geological ground, in particular, has been admirably mapped out by Charles C. Gillispie of Princeton University; but we have valued quite as highly the chance of discussions with such intellectual historians as John Passmore, E. H. Gombrich and Isaiah Berlin. Though in the British academic world today the history of ideas still remains (to use Berlin's pungent phrase) 'a non-subject', we can only hope that our own attempts to reintegrate the ideas of the natural and human sciences may have done something to awaken a greater interest in this field of thought.

As this book is published, the Nuffield Foundation Unit for the History of Ideas is being dispersed, and we ourselves are leaving England for Harvard. During the last four and a half years we have worked on these books and on the associated films under conditions of intellectual patronage such as few academics in Britain ever have the chance of experiencing. As things stand, it appears that educational foundations are the only real means by which novel, exciting—even eccentric—academic ideas can get off the ground. Of all the foundations, the Nuffield Foundation has earned an outstanding reputation for its liberality, flexibility and speed of operation, and we welcome this opportunity to record our deep gratitude to the Trustees for their support. (Too often, alas, when they have played their proper part in initiating novel programmes, the work of units such as ours lapses from the failure of established authorities and institutions to follow up what the foundations began.)

We have dedicated this book to Leslie Farrer-Brown who, as Director of the Nuffield Foundation at the time when we first conceived the idea of this series, took the chance and bore the major responsibility for launching our programme. We are extremely grateful to Miss Caroline Uhlman who typed and prepared this manuscript, and we also owe more than we can say to Miss Helen Mortimer who has played so large a part in all aspects of the Unit's work.

STEPHEN TOULMIN
JUNE GOODFIELD

Alfriston, Sussex
1964

INTRODUCTION

The Problem of Historical Inference

THE picture of the natural world we all take for granted today has one remarkable feature, which cannot be ignored in any study of the ancestry of science: it is a *historical* picture. Not content with achieving intellectual command over the world of their own times, men have been anxious to go further, and discover how the present state of things came to be as it is. Having mapped the existing topography of the heavens and grasped the principles now governing the world of matter, they have also reached back into the darkness of past time, to a period which earlier generations would have found inconceivably remote. Nor have they merely groped in this darkness: on the contrary, they have lit sure pathways, along which the mind can move with confidence. They have learned to do more than guess how the natural world must have looked in earlier epochs; in outline, at any rate, they can now demonstrate how it has in fact changed and developed over vast periods of time. In this way Man has not only succeeded in understanding his environment, but—even more remarkably—has discovered what that environment was like, long before there were any men.

By now, perhaps, we are in danger of underestimating this particular achievement. Once the past is known, it is known; and individual facts about the past are, for the purposes of abstract theory, no more significant than individual facts about the present. Yet, even within science proper, this theoretical standpoint is not the only possible one: for scientific purposes as for any other, the historical point of view can be equally legitimate. During the last century and a half, in fact, speculations about the temporal development of the natural world have begun to mature, so adding a further dimension to our conceptions—the dimension of *time*.

In the whole history of thought no transformation in men's attitude to Nature—in their 'common sense'—has been more profound than the change in perspective brought about by the discovery of the past. Rather than take this discovery for granted, it is almost preferable to exaggerate its significance. One might, for instance, echo Immanuel Kant and ask, 'How is a historical science of Nature *possible at all*?' The Heavens are before our eyes, waiting to be studied. The evidence relevant to planetary theory is ready to hand, so one can scarcely question—except in a metaphysical sense—the possibility of basing astronomy and dynamics rationally upon evidence. It is the same with physics and physiology: atoms and cells may be invisibly minute, and electromagnetic fields intangible, yet they are undeniably present, and can be observed either directly or through their continuing action. But the past is gone, and gone for ever. Anything that can be studied *now* is not (by definition) part of 'the past'; and we can summon present facts about strata, fossils or historical documents to support beliefs about past epochs, only if we make certain assumptions.

We can *demonstrate* how things are now: how they used to be is—if we choose to be sceptical—at best a matter for argument. Indeed, if we carry our scepticism to its philosophical extreme, we may be forced to conclude that, in this particular case, truth is out of our grasp. Let us merely suppose (to borrow an argument from Bertrand Russell) that the whole universe was created five minutes before the present moment, with all the things it contains—our memories included—in their present places and conditions; then every belief about past events remoter than five minutes would be mistaken. If our present evidence is taken in isolation, this may in fact be the case. Our beliefs about Queen Victoria, Julius Caesar, Neanderthal Man and the Ice Ages may be deeply rooted but, compared with our beliefs about present and immediate events, their basis in experience is slender and indirect. Stated in this form, the moral of Russell's sceptical argument may appear somewhat trite; yet his point is surely made. Our knowledge of happenings in the distant past, especially from prehistoric times, is certainly *inferential*. We were not there ourselves, nor do we have the testimony of eye-witnesses. Our beliefs come to us, rather, as the conclusions of reasoned arguments—chains of inference linking data in the present back to the epochs and events in question.

We can now get our two main problems into focus. If the past has gone for ever, how can we be so confident about the course which it has taken? (This we may call 'the problem of historical inference'.) And how did our intellectual picture of Nature acquire its temporal perspective? It is already too late to ask the Kantian question (whether a historical natural science is possible *at all*) for, since A.D. 1800, natural science has widened its scope irreversibly, to embrace historical arguments. So, setting aside the philosophical issues, we can concentrate on more modest questions. How has this change of outlook come about? And how have scientists been able, in half a dozen different directions, to dissipate the mists which earlier enveloped all speculations about the former state of Nature?

Before we embark on our main story, however, we must deal with one preliminary objection. According to R. G. Collingwood, Nature has no history. Only human beings have history. The historian is concerned with the motives and thoughts of those people who have played a part in the events constituting history. Questions of 'motives' do not arise about sticks, stones or brute beasts, but only about creatures who deliberate in a way that we can hope to understand—i.e. about our fellow-men. Though geology, cosmology and evolution-theory may be concerned with temporal development, they cannot (on Collingwood's definition) be called 'historical' sciences.

Yet this particular definition, which deliberately cuts human history off from other enquiries about the past, is useful only up to a point. For, whether we are concerned with the temporal development of the Earth, of organic species, or of some particular society, certain common problems face us in every case. As a matter of logic, we are forced to examine the assumptions implicit in our step back from present-day evidence to the terrestrial, or animal, or human past; and, as a practical matter, human and non-human history share certain methods of enquiry. Archaeological methods now play a large part even within history as most narrowly defined (e.g. in the study of mediaeval villages), and these archaeological methods in turn cannot be sharply distinguished from the methods of geology. For our purposes, the common features of the human and the non-human past are much more significant than their differences. The questions which concern us here affect the perspectives in which we view the past story of Nature and Man alike, embracing the development of the Earth and of living things, as well as of human societies. Partly to emphasize this continuity, partly for lack of a better word, we shall speak here of the *history* of the human species and life on Earth, of the five continents and the globe itself, of the Solar System and the galaxy to which it belongs—and of the cosmos as a whole.

Leaving aside all philosophical issues, educated men now accept certain general propositions. They believe that, before the few thousand years of documented human history, there was a much longer phase of human existence, and that this in turn was preceded by pre-human phases much longer still. Human beings are, moreover, related to earlier forms of life which, if traced far enough back, would lead to a period when simple self-reproducing organisms appeared for the first time. Finally, to go back sufficiently far in time, the Solar System and the galaxies themselves came into their present forms by physical processes whose nature we may yet discover. But scientific enquiry has established, and come to terms with, the enormous extent of past time only very recently. And in the last two and a half centuries, during which the accepted time-scale has expanded from six thousand years to six billion, men have been obliged to re-think all their beliefs, so as to fit them into this new time-perspective. The gradual erosion of long-cherished ideas and convictions accompanying this transformation will be one of our chief concerns in the present book.

What should we make our starting-point? For once this can be chosen without much difficulty. Human beings embarked on the discovery of time with the minds (so to speak) of children. A child, like an animal, lives in the present, and the time-span he can conceive is both restricted and ill-defined. As he grows up, he extends his intellectual and imaginative grasp, and embraces within his span of comprehension his own lifetime, and its relation to the lifetimes of his parents and grandparents. In step with the development of his temporal understanding, the human individual begins to recognize the changeability of his own familiar world and eventually, having accepted the idea of a time before his own birth, he comes to understand that *all* human lives have a beginning, and that once his own parents were infants like himself. So, as the temporal frontiers of his ignorance begin to recede, the first items in the stable background of his world gradually reveal themselves as subject to change. Yet there are limits to his understanding even of this short time-span. How far into the past any individual can reach depends less on his own personal experience than on the cultural and intellectual traditions of his community. Left to himself, he would be capable of going only a couple of generations back. Beyond that, he must rely on memories and hearsay passed down to him by his elders. So the human individual entering on his education—ready for the larger ideas of chronology and historical change which he can acquire only from his society—can symbolize the starting-point of our present investigation.

Do the stages by which we ourselves come to accept a billion-year chronology recapitulate in any way the steps by which earlier generations reached an understanding of this history? Scarcely at all: nowadays, we are introduced directly to the accepted scheme of things, and leap over the centuries of groping and dispute which preceded its establishment. Yet those were centuries during which men saw the whole drama of Nature in quite different terms from ours, catching only occasional glimpses of an earlier state of the world, and scarcely even recognizing those problems which eventually transformed the human vision of Nature. Not having paused to consider the intellectual difficulties which for so long postponed its acceptance, we do not clearly face the question, on what evidence and reasoning our modern time-scale actually rests. (If we did so, we could not possibly take it so completely for granted.) Yet the contrast between the accepted modern picture and a 'child's-eye view' of

Nature could hardly be more extreme. It is one thing to recognize the changes and chances of individual human life—lived on the unchanging stage of the Earth, against the immutable backcloth of the Heavens. But to acknowledge the mutability of the Earth, the living creatures upon it, and even the great Heavens themselves, is something men could do only under the pressure of compelling arguments.

Our agenda must, accordingly, include the following questions. What was the evidence that led men to recognize and admit the mutability of Nature? How could they establish that all those objects and systems, which in the ordinary way seem so stable, were in former times composed and arranged quite differently? How did they work out the time-scales of these changes, and demonstrate the processes which brought them about? Against what alternative and rival conceptions did the modern views have to establish themselves? And how did the arguments actually appear at the time when the issues were still in serious doubt?

The story we tell in the succeeding chapters will answer these questions in broad outline. Unlike the stories of our two earlier volumes, its crucial episodes are extremely recent: even in relation to man's own affairs, the development of a sense of history is still a novel feature of the intellectual tradition. Moreover, almost without exception, the motives, methods and issues involved have been intellectual and religious, rather than practical ones. Neither the needs of modern technology, nor those of the earlier crafts, require one to speculate about the remote past. Wonder, puzzlement, curiosity, imagination—call it what you will: the impulse to pose historical questions has come from within men's minds, and has had little to do with the external demands of life. Nor, for that matter, have historical questions lent themselves directly to experimental study. One cannot experiment on the past: one can only reconstruct it by a kind of scientific detective-work, calling in evidence the results of the experimental sciences when the occasion arises. Thus, at every point, men have been forced to look for analogies between Nature as it was, and natural processes as we find them today. As Lucretius puts it:

> . . . for which cause
> Our age cannot look back to earlier things
> Except where reasoning reveals their traces.

Finally, the scientific points at issue in our present study have constantly been intertwined with religious ones. Astronomy has mapped the Stage on which the drama of Nature is played, matter-theory has listed and set in order the *dramatis personae*; but, once we enter the historical dimension, we begin to be concerned with the Plot. Questions about the History of Nature dovetailed from the very beginning with questions about the religious drama of the cosmos. From our own vantage-point in the twentieth century, the debates about the formation of the Earth and the origin of species appear needlessly violent, bitter and prolonged. But to suppose that, in this field, science and religion could have been kept entirely apart, is to shut one's eyes to the crucial points at issue. If, on occasion, some of the protagonists were perverse, theirs was the desperate perversity of men at the end of their intellectual tether. They preferred to make a last-ditch stand over irrelevant details, rather than surrender their long-cherished presuppositions. For, during the nineteenth century, the History of Nature was rewritten more drastically than ever before or since; and the intellectual choices facing men at that time had unavoidable repercussions on their wider attitudes and beliefs.

I

Memories and Myths

So all Israel were reckoned by genealogies; and, behold, they were written in the Book of the Kings of Israel and Judah.

EACH of us has one immediate and direct link with the past—memory. About some events, we ourselves can speak reliably; about others, we can consult our families and acquaintances. In either case, testimony given in good faith on the basis of clear recollection has a standing all its own, seeming to carry us back and put us in actual touch with the events in question. In our own personal reconstruction of the past, memory is the first essential stepping-stone; we turn to our relatives and neighbours, and through them we are connected with the traditions of our society—that is, with the transmitted recollections of our forefathers. And these traditions are themselves based on memory, quite as much as our own individual recollections: they represent the pooled memories of the community—selected, condensed and handed down from generation to generation.

The immediate ancestor of written history is thus to be found in legend. Indeed, for centuries before the invention of writing (and even after it) men's collective knowledge of the past was based predominantly on what the wisest men in the community said. Initially, exploring the past could mean only one thing: collecting together what was said about past events—the legends—and evaluating them. Those stories which there was reason to trust, one could preserve and repeat: others had to be treated with caution, or dismissed. As for those earlier events which lay beyond the reach of human memory (even transmitted memory), they were beyond the grasp of *knowledge* also. In this way, the scope of memory and legend defined a natural boundary, or 'time-barrier', which set a limit to our possible knowledge of past history. Beyond that time-barrier, men's imaginations entered a new world in which, for lack of trustworthy testimony, they were driven back on to speculation and myth.

In the first and fundamental sense of the term, then, the only testimony we can have about the past is our inherited store of documents and legends. By comparison, all those things which now serve us as indications of the earlier condition of the world—the layers of rock forming the Earth's crust, the rings in fossilized tree-stumps, and the proportions of radioactive carbon in buried ashes—are 'testimony' only in a secondary and transferred sense. For the rocks are just *there*: we can discover the past condition of the Earth only indirectly, arguing back from the condition in which we now find them. They cannot testify for themselves. Their silent gaze has to be interpreted, and we can do this only if we have a well-established theory explaining how they came to their present form.

Now, in the twentieth century, we place as much trust in many historical beliefs for which the sole evidence is indirect, as we do in those for which there is direct human testimony. But in this respect our present century is unique; and, in order to see the origins of our own sense of history in true proportion, we must begin by putting ourselves in the position of our forefathers, who could make no such scientific inferences about the former state of things.

Chronicles and Genealogies

In early societies, then, the place of history was occupied by the collective memory of 'the ancestors'; and the time-scale along which these chronicles were ordered was provided by the names of the ancestors themselves. In a pre-literate or newly literate community, genealogy was not a matter of snobbery or personal curiosity: it was a social art of the first importance, establishing a community's roots in the past, and spinning the lifeline which joined it back to earlier times. All members of the community were linked together by their share in the community's chronicle and family-tree; their standing within the social hierarchy was determined by proof of ancestry, which established their hereditary status; while the prestige and rights of the whole community in the world at large were bound up with its history and antiquity, as certified by the chronicle.

Tribes which possess no written records have, accordingly, strong motives for cultivating the powers of memory. It is the business of the elders of the tribe to keep the inherited histories and genealogies fresh and accurate in their minds, and to pass them on—with any necessary additions—to the rising generation of newly-initiated adults. So on solemn occasions they meet together under the leadership of those with the most tenacious memories, and chant in unison the legends of their forefathers. In this way, they fix the inheritance of oral history more deeply in their minds. To this day, there are communities in Africa whose elders can recite tribal ancestries and chronicles preserving the memory of events going back to the year 1300 or before, while in several countries there still exist illiterate bards who have memorized traditional epic poems, 100,000 lines in length, which they can declaim word-perfect.

Bards and chroniclers are probably related. Some people, indeed, have seen in tribal chronicles one origin of poetry itself. The connection is partly one of metre: when committed to memory and declaimed in unison, traditional stories acquire a natural metrical rhythm, which gives them some of the character of verse. But primitive verse and historical chronicles also share a common function. Tales of an heroic age such as the *Iliad* and the *Odyssey* preserved the national legends of the classical Greeks in the same way that memorized chronicles preserve the records of the African tribes. For the Greeks, Homer was not just a prolific poet, but also—so to speak—their Livy, and even their Moses. As late as the fifth century B.C., in Plato's dialogue, the *Phaedrus*, Socrates laments the spread of reading and writing, on the grounds that it weakens men's powers of memory, and so their critical faculties; and, in Thucydides' lifetime, educated Greeks could still remember when Hellenic history was preserved only in the memories of the people—when the responsibility for transmitting the record of the national past had not yet been shrugged off to books or scrolls.

Compared with many of their contemporaries, the classical Greeks had short memories, and little respect for antiquity or tradition. (This was not simply a deficiency, but in some ways a source of freedom and originality also.) By contrast, the earliest chronicles of the Jews, who formed a closer-knit and more conservative community, had already taken a written form—as the Pentateuch—by about 900 B.C. Writing around A.D. 80, the great Jewish historian Josephus commented sarcastically:

> I cannot but greatly wonder at those who think that we must attend to none but the Greeks as to the most ancient facts, and learn the truth from them only, and that we are not to believe ourselves or other men . . . For you will find that almost all which concerns the Greeks happened not long ago, nay, one may say, is of yesterday and the day before only; I speak of the building of their cities, the inventions of their arts, and the recording of their laws; and as for their care about compiling histories, it is very nearly the last thing they set about. Indeed they admit themselves that it is the Egyptians, the Chaldeans and the Phoenicians (for I will not now include ourselves among those) that have preserved the memory of the most ancient and lasting tradition.
>
> For all these nations inhabit such countries as are least subject to destruction from the climate and atmosphere, and they have also taken especial care to have nothing forgotten of what was done among them,

> but their history was esteemed sacred, and ever written in the public records by men of the greatest wisdom. Whereas ten thousand destructions have afflicted the country which the Greeks inhabit, and blotted out the memory of former actions; so that, ever beginning a new way of living, they supposed each of them that their mode of life originated with themselves. It was also late, and with difficulty, that they came to know the use of letters. For those who would trace their knowledge of letters to the greatest antiquity, boast that they learned them from the Phoenicians and from Cadmus. . . .
>
> [Yet such historians as Cadmus] lived but a short time before the Persian expedition into Greece. Moreover, as to those who first philosophized as to things celestial and divine among the Greeks, as Pherecydes the Syrian, and Pythagoras, and Thales, all with one consent agree, that they learnt what they knew from the Egyptians and Chaldeans, and wrote but little. And these are the things which are supposed to be the oldest of all among the Greeks, and they have much ado to believe that the writings ascribed to those men are genuine.

Following up Josephus' clue, we can reconstruct men's earliest attitudes to the past by looking briefly, not at the legends of the Greeks, but at the sacred records of the great Middle Eastern Empires, which flourished from 3000 B.C. on.

The classical empires of the Middle East were static, hierarchical and conservative. Their social arrangements were tightly organized around an annual cycle of rituals and festivals, which in turn was closely linked to the regular succession of the seasons—notably, to the rise in the level of the great rivers every spring. In both Mesopotamia and Egypt, the focus of political dominion oscillated between the regions near the river-mouths and the inland kingdoms, but these political fluctuations were superficial in their effects. This is especially true of Egypt, where nothing destroyed the underlying continuity of social tradition. At a time when the Greeks retained only confused memories of their history, as preserved in the Homeric epics, the Egyptians had been recording the names of their rulers and the annual fortunes of their kingdoms for many centuries; and these annual records stretched back to well before 3000 B.C.

The most striking relic which has survived of their activities is a series of broken fragments of hard black diorite rock, the best-known of which is the 'Palermo Stone' (Plate 1). These fragments were excavated during the latter half of the nineteenth century, and the chances of archaeology and commerce have scattered them about the world. Originally, they appear to have formed a single slab of rock about two feet high, seven feet long, and rather less than three inches thick: the monument was apparently erected at Memphis, the Egyptian capital, some time before 2500 B.C.—that is, 1250 years before the conjectural date of the Trojan War. A sequence of rectangular spaces was marked out on the face of the slab, and these contained hieroglyphic signs. When deciphered, the first few spaces turned out to contain the names of the earliest Egyptian kings, about whom the collective memory had nothing more to record. At a certain point in the first dynasty (*c.* 2750 B.C.), however, successive rectangles began to record, year by year, the chief events in the chronicles of the kingdom, including the height of the Nile Flood. As time went on, the record became more and more elaborate, even celebrating the launching of ships, the building of new extensions to the King's palace and the capture of prisoners and cattle.

This stone inscription is the oldest body of historical records surviving today. By comparing it with other records written later on papyrus or stone, archaeologists and Egyptologists have, by now, reconstructed a consistent story for the early dynasties of the Egyptian Kingdoms, reaching far back behind the earliest events known to the historians of Greece and Rome—including even their legends.

Such annals and 'king-lists' give, however, a somewhat restricted view of the past. These chronicles may have been treasured as a matter of statecraft, or of sacred obligation, or both; but, from a historical point of view, they had certain serious limitations. They suffered, in part, from the defects common to 'official histories' in any age. Like a roll-call of the Duke of Marlborough's victories, these annals were intended, more often than not, 'to preserve the glory of the present for the future, not to rescue a past from oblivion'. The Kings of the East might, during their lifetimes, subdue their enemies in the flesh, but they could make their power known to later generations only in stone. Even so, there was always the fear that the monuments would be destroyed by their successors—though this fear seems to have been more active in Mesopotamia than in the stabler political environment of Egypt. The most brutal of the Assyrian kings, Ashur-nasir-pal (r. 885–860 B.C.), ended his memorial with an elaborate curse, yet despite his precautions the memory of the Assyrian kings was all but forgotten four centuries later. Shelley's famous sonnet is comment enough on the impotence of these curses—

> . . . And on the pedestal these words appear:
> 'My name is Ozymandias, King of Kings:
> Look on my works, ye Mighty, and despair!'
> Nothing besides remains. Round the decay
> Of that colossal wreck, boundless and bare
> The lone and level sands stretch far away.

The state chronicles of the ancient world had a further defect. For every political centre or grouping had its own chronology: past events were not dated on a common numerical calendar, but referred to the local chronicle of the generations or kings—e.g. 'In the year that King Uzziah died'. In some places, such as Rome, the official chronicles were numbered by the years which had elapsed since the speculative foundation of the city: on our reckoning, 753 B.C. In the Greek city-states, however, historians had at first no way of numbering the years at all, and made do with the lists of dignitaries holding office each year at any city. Only around 350 B.C. did Plato's friend, Timaeus of Sicily, introduce a numerical system for correlating the chronicles of the different cities, dating events by reference to the four-year cycle of Olympic festivals. The astronomers, it is true, could do better. From 747 B.C., a continuous record of solar and lunar eclipses was preserved in Mesopotamia, and this was taken over by Hipparchos of Rhodes, and later by Ptolemy of Alexandria. Ptolemy dovetailed these Oriental astronomical records with his own, so producing the 'Canon of Ptolemy'. With the help of this yardstick, we can date securely events recorded in the Mesopotamian annals from the middle of the eighth century B.C. on, and in many cases can do this to the exact day. Yet the potential value of this eclipse-cycle to history was not immediately recognized. It was not until after A.D. 400 that historians at Alexandria took over the 'Canon of Ptolemy', and related their own hotch-potch of chronological sequences to the common celestial timescale.

In the ancient world, then, the past appeared differently to men of different nations. About the Greeks, Josephus is quite right. They had very little sense of the past and, beyond a few generations, were already plunged into a realm of legend. Even the Trojan War, about 1250 B.C., was a *legendary* event for them. At Memphis and Babylon, the time-scale of the official annals was much longer than any the Greeks knew—though Diodoros of Sicily, writing about 50 B.C., was inclined to be cautious about the alleged antiquity of the Egyptian Kingdom:

> The priests of the Egyptians, reckoning the time from the reign of Helios to the crossing of Alexandria into Asia, say that it was in round numbers 23,000 years. And as their legends say, the most ancient of the gods ruled more than twelve hundred years, and the later ones not less than three hundred. But since this great number of years surpasses belief, some men would maintain that in early times, before the movemen of the sun had yet been recognized, it was customary to reckon the year by a lunar cycle. Consequently, since the year consisted of thirty days, it was not impossible that some men lived twelve hundred 'years'; for in our time, when our year consists of twelve months, not a few men live over one hundred years.

So, not even the most favourably-placed men in the ancient world possessed authentic records going back more than four thousand years; and that figure is the very outside limit. During these centuries, furthermore, the pace of social and political change was never as fast as we are accustomed to today: the very communities with the longest and most reliable chronicles were those in which social and political conditions were slowest to change. Accordingly, there was nothing in men's experience to stimulate any sense of historical development, as we now think of it. Men were aware, of course, of the fluctuations of fortune. Kingdoms rose and fell; but they were always kingdoms of roughly the same kind, with cities resembling their own. Within the limits of their chronicles and communal memories, history could record only the ups-and-downs of chance, trade and war.

The material conditions of their life were even more stable and unchanging. Though geographical changes were undoubtedly taking place in the coastline of the Persian Gulf and the level of the Mediterranean Sea, they were largely insensible. So one finds in antiquity little sense of any continuing development in the geographical and geological scene. On the contrary, most of the men who speculated about the origin of the Earth envisaged it as created at a given moment in time, with substantially its present form and all its major geographical features. What reason had they for envisaging anything else?

Myths and Legend

Yet one important fact remains: they *did* ask questions about the beginning of all things. They could not remain for ever within the temporal span of their surviving documents and legends, either imaginatively or intellectually. Instead, they leapt over the boundaries of this time-span, passing beyond the scope of legend into the cloudier regions beyond—the territory of *myth*. They were themselves not entirely conscious where one region ended and the other began: legends frequently graded into myths, across a hazy borderland in which human (and presumably historical) characters shared the scene with gods and demi-gods. Even in the Homeric epics, the Olympian deities were continually intervening in the action of the Trojan War; and the principal gods of the Egyptians (such as Osiris) had the same double status—part divinities, part legendary heroes. Yet, though in practice the boundary between myth and legend was often blurred, the two ways of describing the past were nonetheless radically different, and the more highly educated Egyptians were evidently aware of this difference. Herodotos visited the temple of Osiris at Thebes in the footsteps of his predecessor, Hecateus of Miletos, and tells us:

> They led me into the inner sanctuary, which is a spacious chamber, and showed me a multitude of colossal statues in wood, which they counted up and found to amount to the exact number [345] they had said; the custom being for every high priest during his lifetime to set up his statue in the temple. As they showed me the figures and reckoned them up, they assured me that each was the son of the one preceding him; and this they repeated throughout the whole line, beginning with the representation of the priest last deceased, and continuing until they had completed the series. When Hecateus, in giving his genealogy, mentioned a god as his sixteenth ancestor, the priests opposed their genealogy to his, going through this list, and refusing to allow that any man was ever born of a god.

The traditional mythologies of the ancient world were intended to provide answers to questions on which no legend, however authentic, could possibly throw light. For suppose one collected together all the traditions recording events for as far back as anyone could recall, and suppose one even called in legends to extend the time-span of the story, the whole result would, nevertheless, be concerned with a sequence of events taking place *within the present dispensation*. Taken by itself, such a story must leave the beginnings of this dispensation unexplained, and could satisfy only those who were content to take the present order entirely for granted. Yet few people were content to do so then, any more than now. A dozen different aspects of Nature and Society cried out for explanation. Communal life, agriculture, the use of wine, fire and metals, and the practice of irrigation: all these had existed throughout the whole of their historical era, yet was it credible that these practices had no origin whatsoever? Somewhere, at the very beginning of things, the necessary arts of social life must surely have been introduced to human beings, or invented by them. Likewise, man's physical environment—the geographical layout of the earth, the pattern and motion of the heavens, and the natural progress of the year: all these were presumably established and set going at some remote epoch of the past.

In this way, men shifted from questions about the succession of events within the present dispensation to questions about its beginning—from 'What has in fact taken place in the world?' to 'How must the present order have originated?' (This was the true boundary between legend and mythology.) If their motives had been only intellectual ones, they might have admitted that no one in fact *could* know about the first origin of things; yet their interest sprang as much from awe and wonder as from curiosity. Faced with the need to keep social life in step with the progress of nature, they could not afford the luxury of complete agnosticism. As we remarked in an earlier volume, ritual and mythology were the twin offspring of social necessity, and their joint origin left its mark on the first creation-stories. So men found it natural to believe that the cosmic order and the social order alike had been established in the beginning through a series of divine actions; and that the arts and crafts on which human welfare depends were taught to men by a series of demi-gods. In every case, the social order was rooted in the Order of Nature: both realms were embraced in a single cosmos, and the annual cycle of rituals duly re-enacted the cosmic drama of Creation, by which the present dispensation was originally set in order.

Measured against our own temporal perspective, the ten thousand years of human society are only a last brief chapter in the development of life on Earth, and represent an even smaller fraction of cosmic history. On such a time-scale, any claim that the laws of society and the laws of Nature are one and the same seems presumptuous. But, in the mouths of the early mythologizers, this claim is more intelligible. For the social arrangements of the Middle Eastern Kingdoms were inextricably bound up with the pattern of the natural environment, so that men were entitled—indeed, compelled—to ask how their own powers were related to the powers of Nature. Nor were they yet in any position to see the early development of the world in any kind of serious proportion. In theory, at any rate, the eras of chronicle and legend had perfectly good time-scales: all the kings referred to lived and reigned for definite periods of time, even if some of these periods were no longer known with accuracy. But, once they crossed the boundary between legend and myth, this last tenuous hold on time was lost. Looking back to the Creation from 1500 B.C., the men of Egypt and Mesopotamia saw all its phases 'end-on'—telescoped together into a single timeless process, within which no clear perspective could be discerned.

Indeed, any inclination to ask about the *dates* of mythological events is probably beside the point. Provided their myths explained the existence of the established order, the question *when* the Creation had taken place was entirely secondary. Clearly, it had occurred before the various events recorded in the surviving legends; but, that apart, it scarcely mattered where in time one located it, or whether one placed it in the temporal sequence of worldly events at all. So, from the very start, the phrase 'In the Beginning' was conveniently vague, and concealed an ambiguity: and, in due course, Aristotle drove a wedge between two senses of the term 'creation'—it could be used in a theological sense, denoting the *dependence* of the world on its maker, or else historically, marking the *first moment* at which the story of the world supposedly began.

At the outset, this ambiguity did not weigh too heavily on men's minds. In most of the early creation-stories, the first steps were in any case somewhat perfunctory: they merely had to set the natural stage on which the more significant, human parts of the drama were to be played. For one thing is certain. These myths did not arise out of disinterested curiosity about Nature: what mattered, rather, was Man's place in Nature. So we find the gods performing their deeds of creation in a landscape recognizably identifiable as ancient Egypt or Mesopotamia. Man's temporal curiosity could scarcely extend any further, for there was no way of pressing questions about the origins of things any distance. While they lacked a general grasp of the workings of Nature, they could hardly form any picture of the past other than the one they did—a few thousand years of gradually increasing obscurity, leading to a point at which the resources of chronicle and legend were finally exhausted. Beyond that lay an impenetrable fog, devoid of any depth or perspective: a 'dream-time', peopled only by the shadowy rumours of unseen mythological powers.

FURTHER READING AND REFERENCES

The earliest attitudes to history have been discussed from a number of different points of view: anthropological, philosophical, historical, theological. For the general historical background, consult

J. H. Breasted: *Ancient Times* and *A History of Egypt*

For an anthropological view of mythology, see the International African Institute's Symposium

African Worlds: ed. Daryll Forde

The best general survey of early historiography is still

J. T. Shotwell: *The History of History*

For specialized essays on a number of different early attitudes to history and time, see

The Idea of History in the Ancient Near East: ed. R. C. Dentan

The general subject of this chapter is touched on interestingly in

F. M. Cornford: *From Religion to Philosophy*
S. H. Hooke: *Babylonian and Assyrian Religion*

For the Palermo Stone, see

J. H. Breasted: *Ancient Records of Egypt*, vol. I

2

Science without History

THE transition from Egypt and Mesopotamia to Ionia and classical Athens for once makes surprisingly little difference. In many fields of scientific enquiry, the methods of thought developed by the Greeks—rational, critical and abstract—set men's minds moving off in new directions. But, when it came to applying their brilliantly original conceptions to explain historical processes in the world of Nature, the Greeks had far less success: in this area they never went, or even claimed to go, seriously beyond the level of intelligent guesswork. Their ideas left them, in fact, no better placed than their predecessors to push back the curtain of the past beyond the limits of memory and legend. For all the rationality of their concepts, they never put down firm intellectual roots into the temporal development of Nature, nor could they grasp the time-scale of Creation with any more certainty than men had done before.

In the History of Nature, therefore, the continuity between the ideas of the Greek philosophers and those of the preceding era is particularly striking: here, even more than elsewhere, one may justly speak of their theories as 'rational myths'. When speculating about the origins of things, they might take care to subject their accounts to logical scrutiny, and to eliminate those personified natural powers which lay at the heart of all earlier myths. But their own accounts of the Creation had only a certain rationality and plausibility to commend them. They were based on little fresh in the way of *evidence*, and they introduced no new *dates*, by which the different stages in the formation of the world could be incorporated into a common time-scale along with the events of human history. At this point, indeed, the Greek scientists fell back on the traditional, tell-tale phrase, 'In the Beginning . . .': for them, the Creation was still something 'flat', telescoped and lacking in time-perspective.

Historical Cosmology in Miletos

This statement must at once be qualified. In the historical sciences, as elsewhere, the first Ionian scientists displayed an initial radical impulse which evaporated later in the more conservative atmosphere of Athens. Somewhere around 550 B.C. the philosophers of Miletos were adopting, for the first time, a consistently developmental attitude to the world. Their view embraced Nature and humanity alike and gave the first hint that the world as we find it is merely one phase in a continuous and continuing creative process. And alongside this first historical cosmology there appeared the first critical historian of human affairs: a man whose ambition was to describe the nations of the world in a way which renounced the mythological traditions and fanciful legends of *all* races and peoples—even those of the Greeks themselves.

The historian was Hecateus of Miletos. He was born into one of the leading families in the greatest centre of international commerce on the Asia Minor coastline. Growing up during the early years of the Persian occupation (from 546 B.C. on), he was a near-contemporary of the philosopher, Anaximenes. He was lucky in the time and place of his birth. The Ionian school of philosophers was already in its third generation, and the sudden expansion of the Persian Empire had opened up new areas of exploration, to the east and the south. Hecateus made the most of his chances. He set off on the Grand Tour, in much the same spirit as a wealthy young European of the eighteenth century: his encounter with the Egyptian priests at Thebes knocked the last of his Greek parochialism out of him, and he was quickly forced to recognize that the historical traditions current in the various nations were all equally inadequate and open to question. Returning home, he settled down and produced two books, each of which was, in its own way, an early landmark in the development of historical thought. One of these, *Travels Round the World*, introduced his contemporaries to 'comparative ethnology'—deliberately bringing home to them the variety in the customs and the inconsistency of the beliefs current in different parts of the world. In his other book, the *Genealogies*, he passed on to his fellow-citizens the lesson which he himself had learned at Thebes. It opened with a brave flourish—

> Hecateus of Miletos declares as follows: I am writing only what I myself consider to be true, for the traditional tales current among the Greeks are manifold, and strike me as ridiculous—

and went on to examine, with a new scepticism, the accepted legends about the early history of Miletos. By the standards of its time, Hecateus' travel-book immediately became a best-seller. His descriptions of Persia and Egypt, his reports on the fauna, flora, customs and legends of the countries he visited, were repeated again and again by later historians and compilers, and their influence can be detected as late as 50 B.C., in the pages of Diodoros of Sicily.

The opening chapters of Diodoros' *Historical Library* reflect not only the historical ideas of the Ionians but also their cosmology, and reproduce a creation-story much of which probably originated with Anaximander. The first part of this describes the process of creation up to the appearance of the higher animals:

> When in the beginning, as their account runs, the universe was being formed, both heaven and earth were indistinguishable in appearance, since their elements were intermingled: then, when their bodies separated from one another, the universe took on in all its parts the ordered form in which it is now seen; the air set up a continual motion, and the fiery element in it gathered into the highest regions, since anything of such a nature moves upward by reason of its lightness (and it is for this reason that the sun and the multitude of other stars became involved in the universal whirl); while all that was mud-like and thick and contained an admixture of moisture sank because of its weight into one place; and as this continually turned about upon itself and became compressed, out of the wet it formed the sea, and out of what was firmer, the land, which was like potter's clay and entirely soft. But as the sun's fire shone upon the land, it first of all became firm, and then, since its surface was in a ferment because of the warmth, portions of the wet swelled up in masses in many places, and in these pustules covered with delicate membranes made their appearance. Such a phenomenon can be seen even yet in swamps and marshy places whenever, the ground having become cold, the air suddenly and without any gradual change becomes intensely warm. And while the wet was being impregnated with life by reason of the warmth in the manner described, by night the living things forthwith received their nourishment from the mist that fell from the enveloping air, and by day were made solid by the intense heat; and finally, when the embryos had attained their full development and the membranes had been thoroughly heated and broken open, there was produced every form of animal life. Of these, such as had partaken of the most warmth set off to the higher regions, having become winged, and such as retained an earthy consistency came to be numbered in the class of creeping things and of the other land animals, while those whose composition partook the most of the wet element gathered into the region congenial to them, receiving the name of water animals. And since the earth constantly grew more solid through the action of the sun's fire and of the winds, it was finally no longer able to generate any of the larger animals, but each kind of living creatures was now begotten by breeding with one another.

As so often with the early Greek scientists, this account is remarkable less for its details than for its general form. Its most significant point is this: the world of Nature is supposed to have acquired its present form as the outcome of a prolonged sequence of gradual changes, all of which the Ionians could still observe in some form or other. There is no reference to arbitrary acts of Divine command, but only to the behaviour of whirling eddies, potter's clay, the warm moisture of swamps and marshes continually generating living things, and above all the gradual drying-out of the marshes, as the land slowly but inexorably encroaches upon the sea. To citizens of Miletos this image was particularly forcible, for the coastline of Ionia was advancing year by year into the Aegean. The harbours of Miletos were silting up, and the marshlands of the River Meander, at whose mouth the city lay, were slowly but surely turning into cultivable land. (Now in the twentieth century, the site of Miletos lies several miles inland, and the whole sweeping arm of the sea which it formerly commanded has been turned into orchards, cornfields and sheep-pastures.)

In any historical cosmology, the question of the origin of life is inescapable. Anaximander envisaged the first generation of living things as coming into being through a metamorphosis, like that which transforms a chrysalis into an insect. The chrysalis stage, corresponding to Diodoros' 'delicate membranes', was represented by a 'fish-like' state, which recalls the form of sea-urchins:

> The first animals were generated in the moisture, and were enclosed within spiny barks. As they grew older, they migrated onto the drier land; and, once their bark was split and shed, they survived for a short time in the new mode of existence.
>
> Man to begin with was generated from living things of another kind, since, whereas others can quickly hunt for their own food, men alone require prolonged nursing. If he had been like that in the beginning, he would never have survived. . . . Thus men were formed within these [fish-like creatures] and remained within them like embryos until they had reached maturity. Then at last the creatures burst open, and out of them came men and women who were already able to fend for themselves.

Some readers have taken these surviving fragments from Anaximander as a brilliant anticipation of nineteenth-century ideas about evolution. But this is to give him credit for quite irrelevant reasons. In fact, he did even less to anticipate Darwin than Demokritos did to anticipate Dalton. His account in no way implied that living species can change. It merely explained how the first individuals of each particular species came into existence without parents of their own kind. (The egg came before the hen.) The primary 'fish-like creatures' were not organisms of a more primitive species, but only the special embryos required for the safe development of this first generation. The living creatures produced in this way were *already* divided up into the different species of land, air and water-animals, and they continued to propagate their own kind in fixed species, by inter-breeding among themselves. The outstanding originality of Anaximander's cosmology is not any foreshadowing of Darwinism: it is the fact that he portrays the creation as a gradual and continuous—i.e. as a *historical*—process.

In point of fact, the zoological phase in Anaximander's creation-story involves its only discontinuity. During classical times, nothing in men's experience suggested that species could develop one out of another, and even the most radical philosophers of antiquity never envisaged the possibility of organic evolution. To all the classical philosophers, the phrase 'the origin of life' meant the natural processes (if any) by which living things came into existence *in their present species.*

The Ionian cosmology also explained the beginnings of society, language and the social arts. As Diodoros reports—

> But the first men to be born, they say, led an undisciplined and bestial life, setting out one by one to secure their sustenance and taking for their food both the tenderest herbs and the fruits of wild trees. Then, since they were attacked by the wild beasts, they came to each other's aid, being instructed by expediency, and when gathered together in this way by reason of their fear, they gradually came to recognize their mutual characteristics. And though the sounds which they made were at first unintelligible and indistinct, yet gradually they came to give articulation to their speech, and by agreeing with one another upon symbols for each thing which presented itself to them, made known among themselves the significance which was to be attached to each term. But since groups of this kind arose over every part of the inhabited world, not all men had the same language, inasmuch as every group organized the elements of its speech by mere chance. This is the explanation of the present existence of every conceivable kind of language, and, furthermore, out of these first groups to be formed came all the original nations of the world.
>
> Now the first men, since none of the things useful for life had yet been discovered, led a wretched existence, having no clothing to cover them, knowing not the use of dwelling and fire, and also being totally ignorant of cultivated food. For since they were ignorant of the harvesting of the wild food, they laid by no store of its fruits against their needs; consequently large numbers of them perished in the winters because of the cold and the lack of food. Little by little, however,

> experience taught them both to take to the caves in winter and to store such fruits as could be preserved. And when they had become acquainted with fire and other useful things, the arts also and whatever else is capable of furthering man's social life were gradually discovered. Indeed, speaking generally, in all things it was necessity itself that became man's teacher, supplying in appropriate fashion instruction in every matter to a creature which was well endowed by nature and had, as its assistants for every purpose, hands and speech and sagacity of mind.

So much for the 'prehistoric' origins of society: having summarized one of the more plausible current creation-stories, Diodoros could turn to the main business in hand, and collect together, into a full historical account, 'all the events handed down by memory and taking place in known parts of the inhabited world'.

Despite its brevity, this speculative account is once again original and level-headed, looking forward to a future science of anthropology rather than backwards into earlier mythology. For society, as for nature, the keynote of the Ionian story was gradualness and continuity—'bit by bit', 'little by little'. There was no indication, of course, just how long men had lived like wild animals, or just how recently they had learned the use of fire. But the Ionians did at any rate envisage some sort of temporal perspective, beginning with that remote era when the higher animals came into existence, and continuing through a sequence of prehistoric epochs when the social arts were slowly and painfully acquired.

The Divorce of History from Philosophy

So, for a short time in the sixth century B.C., there was the prospect of an alliance between history and natural philosophy which might have left an historical stamp on all succeeding scientific thought. But things did not work out in this way. Historians and scientists soon parted company, and the dominant influence on Greek natural philosophy came, not from history, but from mathematics. It was a case of divorce by mutual consent. Both historians and philosophers found reasons to concentrate on more immediate problems, turning their attention away from the shared spectacle of Nature and Society evolving together. We in the twentieth century can understand and sympathize with the historians' reasons, for the pressure of great events was turning them into *contemporary* historians. Writing before 500 B.C., Hecateus could survey the nations of the world with an Olympian eye, and justify his cosmopolitan scepticism by appealing to the experience of his travels. The next half-century, however, saw the Persian assault on the mainland of Greece, culminating in the repulse of Xerxes' forces after Marathon and Salamis. With these dramatic battles, the tide of history seemed to have turned; and, writing as he was for the Athenians who had actually taken part in them, Herodotos of Halicarnassos could scarcely take up the same detached standpoint. (One might as well expect a Russian historian of the 1950s to approach the campaigns of Hitler in the spirit of a scientific anthropologist.)

Yet Herodotos did incorporate into his history of the Persian War a great deal of miscellaneous background material. Evidently his readers were avid for information, and much of the fascination of his work still comes from his continual digressions. He includes, for instance, a rambling survey of the history and legends of the ancient empires, compiled from a variety of sources, including Hecateus. Two things, however, mark him off as a real historian. All his digressions are attached more or less tenuously to the contemporary historical narrative which is his central theme; and, in handling second- and third-hand sources, he is careful and circumspect. He questions, for instance, whether the Phoenician expedition despatched to circumnavigate Africa could really have reached a point at which the Sun crossed the sky to the north, and he regrets that he has met no eye-witnesses able to confirm the existence of a Northern Sea on the farther side of Europe, and of the 'Tin Islands' (perhaps the British Isles).

A generation later, these last threads linking history to natural science had been broken. Herodotos' successor, Thucydides, had no interest whatever in the remote past, and dismissed it summarily in his first chapter:

> Judging from the evidence which I am able to trust after most careful enquiry, I should imagine that former ages were not great, either in their wars or in anything else.

Thucydides himself was plunged into what he believed to be the greatest war in history—the struggle between Athens and Sparta known as the Peloponnesian War. If Herodotos wrote the history of a period which was still within living memory, Thucydides went one better: he collected his material as the events took place, beginning with the outbreak of war in 431 B.C. The twenty-seven years of fluctuating hostilities occupied the prime of his life, and in his opinion even the Persian War was a second-rate affair by comparison. For the characters and aptitudes of men and political systems were being tested by the Peloponnesian War as never before, and Thucydides set out to demonstrate how the virtues and vices of the opposing parties played their parts as the final instruments of destiny.

This contemporary, Churchillian theme gave Thucydides' work its characteristic merits and defects. On the one hand, it had actuality, psychological interest, and a dramatic insight into the ideals and appetites which are the immediate driving-force of politics. On the other hand, he failed to view the conflict between the Athenians and Spartans in a wider context, he was blind to the economic and social changes which had been its prime conditions, and he treated human nature too narrowly—as something entirely fixed, shaping the course of history without itself being shaped in return. With Thucydides, in short, Human History had become quite detached from the History of Nature. We miss the earlier sense of a continuous and continuing historical development, linking the present state of society back to a primitive, prehistoric existence. From this time on, the subject-matter of history embraced the present era, together with its immediate antecedents; so that, even with seven hundred years of Roman history to play with, Livy could treat Rome as Thucydides had done 'human nature'—as an unchanging entity. The sense of evolutionary development, temporarily awoken by Anaximander and Hecateus, had been lost.

If by 400 B.C. the attention of historians had become focussed on contemporary problems, so too had the attention of philosophers, but in a different way. For the problems which preoccupied Plato and his successors at Athens originated as much in Italy as in Ionia—coming from Parmenides of Elea, and from the followers of Pythagoras. Their questions slanted the philosophical debate in a new, metaphysical direction, which was far more significant for mathematics and theoretical physics than for the historical aspects of natural science. As time went on, almost everything in philosophy became subordinated to the 'problem of change'—of explaining the 'transitory flux' of experience in terms of the 'unchanging realities' that lay behind it. Once the axiom was accepted, that all temporal changes observed by the senses were merely permutations and combinations of 'eternal principles', the historical sequence of events (which formed a part of the 'flux') lost all fundamental significance. It became interesting only to the extent that it offered clues to the nature of the enduring realities. So questions about historical change ceased to have any relevance to the central problems of philosophy, and philosophers concerned themselves instead with matters of *general principle*—the geometric layout of the heavens, the mathematical forms associated with the different material elements, or the fundamental axioms of morals and politics. More and more, they became obsessed with the idea of a changeless universal order, or 'cosmos': the eternal and unending scheme of Nature—society included—whose basic principles it was their particular task to discover.

Guided by Pythagoras, the philosophers of the Academy looked for these principles in the realm of mathematics. Whatever could not be interpreted in geometrical terms ceased to be of profound concern: like a quantum theorist contemplating meteorology today, they saw in the historical flux only a tedious and elusive muddle. All the leading Athenian philosophers, even those (like Aristotle) who refused to base all scientific theory on mathematics, undervalued the temporal flux, and were accordingly distracted from the problems of history. Outside the history of philosophy, Aristotle made only one excursion into history, the introduction to his *Constitutions of Athens*, a perfunctory affair which may well have been ghost-written by one of his students. In general, he had no use for speculations about origins, on the grounds that they were unverifiable: no evidence being available either way, he preferred to concentrate on those things—the planetary system, human virtues and vices, the structure of society, and the life-cycles of organisms—which were there before his eyes, readily available for reflective study. The Epicureans apart, the principal schools of philosophy from the fourth century onward continued to treat the temporal flux of events as irrational and misleading.

What, then, could they say about the Creation? Having rejected the cosmological approach of the Ionians, they forfeited its two chief advantages. For Anaximander's History of Nature had been both empirical and 'uniformitarian'. Though he based his account on intelligent guesswork rather than on solid evidence, he did introduce into it a number of elements which in due course might be determined empirically: e.g. the actual lengths of those prehistoric phases by which men moved from a brutish existence to settled city-life. Secondly, on his account the Creation was continuous with analogous processes still going on, so that—in theory at least—one could answer questions about the past by arguing back from present evidence.

These were the advantages that the Athenian philosophers lost. If the Order of Nature was 'eternal and unchanging', there could be no question of its having come into existence progressively, and continuously. On the contrary, if there had been a beginning in time at all, then the universe must have been created at one stroke, already complete in every essential feature, and with substantially its present form. Nor could the course of this Creation ever be discovered by studying the changes going on today. For there was no analogy whatever between the creative phase of cosmic history, during which the Order of Nature was *instituted*, and the subsequent course of events which was a *consequence* of this order. One could therefore establish the character of the Creation, only by arguing from first principles. This did not worry Plato and his successors at all. In certain respects, indeed, they found the conclusion a welcome one: to have based one's cosmology on empirical data, belonging to the flux, would have started a rot in the foundations of their whole intellectual system.

So, for philosophical purposes, the Creation became detached from the subsequent development of Nature, and the origin of the cosmos became a topic for abstract argument. Over the resulting problems, each of the four main schools of philosophers—Platonists, Aristotelians, Stoics and Epicureans—adopted its own characteristic position, and their doctrines have largely shaped subsequent attitudes to History and Time. We must briefly consider each school in turn.

Plato's Creator-Craftsman

For Plato and his disciples, the fundamental problem arose in the following way. The World of Nature formed an ordered system, with a permanent, mathematical character, which governed also the social and moral orders. If this mathematical order had any source and origin, it could not be due to a continuing material process, as envisaged by the Ionians. The question was, what alternative account could one give? One was forced back, Plato concluded, into a frankly mythical story.

Plato's creation-myth, preserved for us in the *Timaeus*, takes as its starting-point an absolute contrast between mathematical formulae, which are eternally valid, and the temporal events of the world, which are in constant flux. The Creation of the cosmos was the process by which the eternal mathematical principles were given material embodiment, imposing an order on the formless raw materials of the world, and setting them working according to ideal specifications. A 'creation', understood in this highly sophisticated sense, could hardly be described in familiar everyday terms, and Plato resorted—quite deliberately—to mythological language, using the analogy of a Divine Craftsman as his figure of speech.

The Craftsman's task is, first, to conceive the ideal specification to which Nature is to conform, and then—so far as the nature of the raw materials permits—to shape, arrange and set them in motion in conformity with this specification. In modern terms, the Craftsman inaugurates the existing 'laws of nature': the intellectual programme of modern mathematical physics, as conceived in the seventeenth century, being one with which Plato would have sympathized. (Galileo and Newton were as concerned as the philosophers of the Academy to reveal the blueprints of the Creator—the 'Intelligent Agent' of Newton's natural philosophy, 'through whose Counsel all material Things were composed'.)

If Plato's story were genuinely historical, we should at once want to ask: 'In what sequence did the Craftsman perform his creative acts? And how long ago did this all take place?' Its mythical character shows itself in the fact that neither question is of any relevance. The *order* of the Creator's agenda was of no importance, because Plato was only concerned to prove the rational character of his end-product, and demonstrate the imaginative conception underlying its present structure. Given his passion for understanding, one can see the point of his preoccupation. The heart of a motor-car factory (for instance) lies not in the assembly-shop, but rather in the offices of the designers who conceive and plan the cars. Only when we have mastered the designs can we understand all the other things going on in the factory: until this is done, the mechanical processes which give a material embodiment to the designer's conceptions will remain unintelligible. As Plato saw the problem, the same was true of the entire universe. When a mathematician worked out the geometrical relations manifested in the motions of the planets, or the properties of material substances, he was likewise revealing the Craftsman's rational design.

Plato was equally non-committal about the *timing* of the supposed Creation: it was not the sort of event whose date could be inferred from the evidence of its after-effects. Certainly, if one was to give it a date, it must have occurred more than nine thousand years back. (This was the period of time which, on his estimate, had elapsed since the legendary struggle between the Athenians and the men of Atlantis; and this in turn was only one incident in a series of historical cycles, by which civilizations had repeatedly risen and fallen.) If he had been pressed, Plato would probably have swept aside as absolutely irrelevant all questions about the date of the Creation. For here, in the *Timaeus*, one central ambiguity was already becoming evident, and this was to reappear in later Christian doctrines. To call God 'the Creator of the World' may be to say that, at some moment in time, he brought it into existence; alternatively, it may mean that the world is, now and at all times, dependent on God for its continued existence.

Aristotle, polemical as always, exploited this ambiguity to attack Plato's theory. He interpreted the *Timaeus* quite literally, as implying that the world came into existence at some definite moment in time, and went on to argue that this view landed one in inconsistencies. Yet Plato's pupils rejected this interpretation as unfair, replying that the Creation was a kind of intellectual construction, which explained the foundations of order in Nature, not its beginning.

Plato's preoccupation with ideals rather than actualities also underlies his off-hand attitude towards the history and mechanics of social change. The principles of the social order were, to his way of thinking, as eternal and unchanging as the principles of the natural order. States were happy and well run to the extent that they conformed to this ideal pattern. These ideals could, no doubt, be subverted in a hundred different ways, but this fact was hardly important: the statesman's task was to reflect on the principles of political health, not to tabulate all the possible varieties of social affliction and disease. In practice, no doubt, human fortunes had fluctuated, so providing the subject-matter for history. At some epochs, men had slowly improved their crafts and conditions of life: at other times, natural catastrophes or human stupidity destroyed men's cities and threw them back to the level of subsistence.

But the changes and chances of history did not invalidate the fundamental principles of morals and politics: they were merely irrelevant. The idea that the 'principles of a healthy social order' might themselves vary from age to age, and from country to country, was one that Plato could never accept. For in society, as in Nature, there existed a permanent body of unchanging truths, by which a wisely governed city should be guided at any place or time. There was thus no motive for a Platonic philosopher to study the temporal flux of human history. Apart from serving to illustrate the principles of political philosophy, what else could it possibly teach him?

Aristotle's Eternal Universe

In reconstructing earlier attitudes towards the history of the natural world, we are brought up short against one question: What constitutes 'evidence' in science? This question arises for us here in two ways. To begin with, many things which now serve us as evidence of the Earth's early development were not available as such for the Athenians—just because they were not recognized *as being* evidence, so that their *relevance* was not appreciated. Secondly, the small amount of genuine evidence actually available gave only a partial and limited insight. However far the Greek philosophers agreed on the facts of the case, they could without inconsistency interpret them in drastically different ways. As a result, though he started from the same data as Plato, Aristotle ended up by viewing historical events and sequences in a very different aspect and proportion.

His starting-point was extremely close to Plato's. If anything, Aristotle insisted even more strongly that the philosopher must account for the *present* Order of Nature—building up a system of concepts which does justice to the world as it now exists, rather than indulging in speculations about some long-past, and entirely hypothetical, *Genesis*. By restating the problem in these terms, Plato had done philosophy a service. One must begin from the existing Order of Nature, and the fundamental question at issue was not *when*, but *how* and *why* it was so ordered. In any case, information about a supposed 'beginning of all things' did not help one to understand the how and why of the cosmos. There is more to a fully-developed animal than can be discovered by looking at the embryo from which it comes; and why should things be any different for the cosmos as a whole?

Furthermore, as Aristotle pointed out, there were good reasons for supposing that there had never been such a 'beginning of all things', and that the natural order was eternal and unchanging, having neither a temporal origin in the past, nor any prospect of destruction in the future. He defended this conviction with a variety of considerations—observational, logical and religious; but his deepest reasons were metaphysical ones, and those sprang from his fundamental disagreement with Plato's system. The stable patterns in terms of which the structure and workings of the natural world were to be interpreted should not be regarded as abstract mathematical ideals: for Aristotle, they were matters of actual fact. The principles of the natural order had to do, not with some detached world of intellectual forms, but rather with the actual behaviour and properties of tangible material things.

Logic, observation and piety all reinforced this initial position. For Aristotle, there was something grotesquely impious in the idea that the entire universe was just another mortal creature; while the conception of an entire universe coming into existence out of nothing, at some definite instant, created needless logical quandaries about the idea of time. In the *Timaeus*, Plato had apparently been attempting to bring 'time itself' into existence at some moment *in time*; yet how could one even speak of 'the moment when time began', without immediately prompting questions about what happened *before* that? The concept of time (Aristotle argued) was essentially *unbounded*, and inconsistent with the idea of a 'moment' of creation. (This difficulty was to recur, centuries later, for Nicholas of Cusa, Immanuel Kant, and the physical cosmologists of the twentieth century.) Observation, in turn, supported logic and metaphysics; for what was there in our experience to suggest that the natural order was anything but eternal? The inexorable revolutions of the heavenly bodies, the remarkable stability of organic species, the ups-and-downs of human fortune and the seeming cycles of geological change: in all these things, he could find (as James Hutton was to put it much later) 'no Vestige of a Beginning—no Prospect of an End'.

Aristotle's belief in the fixity of organic species was, accordingly, based on something far more profound than inadequate biological information. Nor could this view have been countered by appeal to any evidence then available. Even at the narrowly biological level, all the known facts supported it. Through the constancy of the species, indeed, each individual could share something of the perfection and immortality with which those celestial divinities, the stars, were blessed. 'Things that are born and die imitate those that are imperishable': organic change followed a rhythm as regular as the rhythm of the heavenly movements, and harmonized with it.

Even the rise and fall of civilizations might perhaps conform to the same overall rhythm. In this connection, both Aristotle and Plato toyed with an attractive and sweeping hypothesis. Once every few thousand years, the Sun, Moon and planets returned to the same relative positions, and began to follow out again the same sequence of configurations; so perhaps the rhythm of political fortunes also had its own definite period, keeping the recurring cycles of social change in step with the motion of the Heavens. If that were so (Aristotle remarked) then he himself was living *before* the Fall of Troy quite as much as *after* it; since, when the wheel of fortune had turned through another cycle, the Trojan War would be re-enacted and Troy would fall again.

Stoics and Epicureans

Despite Aristotle's exhortations, later Greek philosophers did not renounce speculations about the origins and history of the cosmos. If anything, their curiosity actually intensified after Aristotle's death, as they became more and more sceptical about the traditional gods of Olympus. It is true that only a small fraction of the population took any direct interest in these philosophical doctrines—educated men by and large shrugged them off with a laugh, in the same way that early twentieth-century England shrugged off the paradoxes of Bernard Shaw. But, as Shaw knew, the power of comedy can often outflank mental obstacles which resist direct assault. Even those who laughed at Aristophanes' caricature of Socrates came to realize that the deities of the traditional mythology could no longer be accepted light-heartedly and uncritically. This scepticism, in turn, stimulated the development of rival philosophical theologies, whose aim was to reinterpret and even to supersede the older Hellenic myths.

Of the two most influential schools of thought, it was the Epicureans who were the more radical sceptics, explaining away religious fears and superstitions as irrational and unnecessary. The Stoics, by contrast, kept the older religious ideas in circulation, but tried to bring them up to date by 'de-mythologizing' them. Cicero summed up their position as follows:

> Chrysippus, who is look'd upon as the most subtle interpreter of the dreams of the Stoics . . . says that the divine efficacy is placed in reason, and in the spirit and mind of universal nature; that the world, with a universal effusion of its spirit, is God . . . He maintains the sky to be what men call Jupiter [Zeus]; the air, which pervades the sea, to be Neptune [Poseidon]; and the earth Ceres [Demeter]. In like manner he applies the names of the other deities. He says that Jupiter is that immutable and eternal law, which guides and directs us in our manners; and this he calls fatal necessity, the everlasting verity of future events.

In this way, theology was subordinated to cosmology, and the cosmology of the Stoics was closely bound up with their theories about the material composition of the world. (On these theories, see *The Architecture of Matter*.) They too were convinced that the system of Nature was based upon rational principles—even that the objects of the world were all bound together in a determinate web of actions and reactions. This accounts for the streak of fatalism associated with their moral system, which we hint at today when we describe a man as 'stoical'. But they did not share Aristotle's further belief that the fundamental layout of Nature must be stable and unchanging. On the contrary, they carried farther the idea of a 'cosmic rhythm', by applying it to the development of the whole universe, and ended by producing a cosmology in which everything was caught up in one common, all-embracing cycle of death and rebirth. In this system, Heavens and Earth, gods and men, were all fated to act as they did, so even the traditional deities could no longer be suspected of wilful malice.

The Epicureans took the process of demythologizing still further, and produced a naturalistic cosmology, from which the gods were rigidly excluded. Though their account of the origins of Nature and society was in some ways less well argued than those of their rivals, it had two outstanding merits. It was rich in detail, and it presented the successive phases of the Creation historically, in a plausible order and perspective. As expounded in Lucretius' poem *On the Nature of Things*, the Epicurean account covered in succession the formation of the Earth and the planets, the origin of life, and the rise of society. For our purposes the most important part of the whole system is Lucretius' account of the origin of life, for this put into general circulation ideas whose influence on European thought has been far wider than that of Epicureanism itself. Like Hesiod long before him, Lucretius believed in a Golden Age—an earlier and more blessed phase of history, when Nature was more productive, men were stronger, and life was easier. Whereas in the present epoch only the lower animals were spontaneously generated, in the Golden Age this was true for living creatures of all kinds—even human beings. At that time the soil was more fruitful, the climate was better, and the Earth secreted from itself a kind of nutrient 'milk': as a result, the first human infants were able to survive to maturity without adult nurses.

> Earth to these [first] children furnished food, the heat
> Clothing, the grass a bed well lined with rich
> Luxuriance of soft down. Moreover then
> The world in its fresh newness would give rise
> Neither to rigorous colds nor excessive heats
> Nor violent storms of wind.

Yet one thing about Lucretius' theory must be emphasized. He believed that living things came into existence at some definite moment in the past, and that they did so by perfectly natural processes; he believed, too, in a natural selection; yet, nevertheless, he rejected outright any suggestion that organic species could change or evolve. Indeed, he spends nearly a hundred lines of his poem refuting this possibility.

> Centaurs there have never been, nor yet
> Ever can exist things of two-fold natures
> And double body, moulded into one
> From limbs of alien kind . . .
> *But each thing has its own process of growth;*
> *All must preserve their mutual differences,*
> *Governed by Nature's irreversible law.*

The function of natural selection was not to change the species, but simply to eliminate monsters and weaklings thrown up spontaneously at the beginning:

> Such monstrous prodigies did Earth
> Produce, in vain, since Nature banned their growth,
> Nor could they reach the coveted flower of age,
> Nor find food, nor be joined in bonds of love;
> For we see numerous conditions first
> Must meet together, before living things
> Can beget and perpetuate their kind . . .
> And many breeds of animals in those days
> Must have died out, being powerless by their offspring
> To perpetuate their kind. For all those creatures
> Which now you see breathing the breath of life,
> 'Tis either cunning, or courage, or again
> Swiftness of movement, that from its origin
> Must have protected and preserved each race.

Finally, he speculates about the beginnings of social life, language, religion and the crafts. His account, like that of Diodoros, looks back to the Ionians—necessity and the increasingly harsh conditions of life driving men to one innovation after another. Or so, at any rate, it *seemed* . . . ; for, after devoting more than a thousand lines to his creation-story, Lucretius candidly admits that the whole thing has been nothing more than a reasoned fable, lacking a solid foundation of evidence, and justified only by inference and guesswork:

Now men dwelt securely fenced around
By strong powers, and the land was portioned out
And marked off to be tilled; already now
The sea was gay with flitting sails, and towns
Were joined in league of friendship and alliance,
When first poets made record in their songs
Of men's deeds: for not long before this time
Had letters been invented. *For which cause*
Our age cannot look back to earlier things
Except where reasoning reveals their traces.

The Limits of the Classical World-Picture

At this point, Lucretius puts his finger on the central difficulty in all attempts to theorize about the past. Beyond the boundaries of chronicle and legend, everything depends on the speculative interpretation of indirect evidence; and this evidence has to be not only available, but recognized for what it is. For the Greek natural philosophers the difficulty was particularly acute, and their creation-stories could be little more than reasoned fables, or rational myths. Until some solid evidence was found to indicate what had in fact happened in the remote past, men could do no more than ask how nature might have developed, and so sketch the outlines of possible cosmologies. Even at the beginning of the Christian era, with the great period of ancient scientific thought behind them, the men of Alexandria and Rome knew no more than their predecessors of six centuries before about the history of Nature, or the remoter history of the human race . . . and that was very little.

So, at the end of the classical period, men's vision of Nature remained, fundamentally, as unhistorical as it had been at the beginning. No way could yet be seen to break through the time-barrier which confronted them at the point where the evidence of documents and legends finally gave out. Until some rational principles had been established to control arguments back to prehistoric times—until some reliable modes of inference had been worked out, by which the time-barrier could be outflanked or overleaped—ideas about the origin and development of Nature during prehistoric times were necessarily speculative. Speculation apart, the ancients were confined within the limits of their own epoch: fog-bound (as it were) in a stable world, whose basic structure—for all that they could see—had always been what it was in their own day. At best, the Greek scientists could thin this fog only locally and temporarily, since the theoretical possibilities they imagined far outran the facts available to them as evidence.

Still, later generations were indebted to them for two things. In the first place, they had isolated the crucial problems involved in any attempt to reconstruct the past history of the natural world and the origins of social life; and secondly, they had experimented with almost all the forms of theory which were to dominate later thought, and had recognized the characteristic merits and defects of each. In Plato's *Timaeus*, we have the nearest thing in pre-Christian philosophy to a 'Big-Bang' cosmology, in Aristotle the outlines of a 'Steady-State' theory: the Stoics, in turn, pioneered a 'Cyclical Cosmos', while the Epicureans saw the development of the world rather as a random, One-Way Process. Not until many centuries later could scientists put the history of Nature on a solid basis of established facts and secure inferences. But, in the intervening time, the patterns of thought introduced by the philosophers of classical Greece kept their hold on the human imagination; and, when men were finally able to extend their rational grasp to include the remoter past, they turned once again (as we shall see) to these original forms of theory for their intellectual scaffolding.

The Beginnings of Natural History

Meanwhile, the Greek natural philosophers were perfectly aware of those phenomena which, during the last 150 years, we ourselves have recognized as providing clues about the remote past. Yet their whole approach to the natural world tended to discourage an historical interpretation of Nature, and so, though they saw the same beasts and birds, rocks and rivers as we do ourselves, they interpreted their forms and behaviour quite differently. To begin with, this was no obstacle: as Aristotle and his followers demonstrated, biology can go a long way before any evolutionary questions arise. The fundamental conviction that different organic species are permanent elements of the natural order had a double effect. On the one hand, it added a philosophical stimulus to curiosity about living things; but, at the same time, this attitude inevitably imposed a static scheme of interpretation, and for nearly two thousand years there was no serious reason to call in question either the premise or the conclusion.

Aristotle, his immediate pupils and his correspondents together formed one of the great generations in science, to which all subsequent natural historians owe a debt. In his own lifetime, Aristotle's network of pupils and informants sent back from all parts of the known world a continuous stream of material for his encyclopaedias of zoology, comparative anatomy and physiology. Theophrastos, perhaps the most distinguished individual of them all, took over direction of the school after Aristotle's death, and wrote the first fundamental accounts of the structure and classification of plants. The resulting books were so fully documented, so profound, and so embracing in their scope, that they provided the groundwork of biological research for many centuries. They were superseded only in the seventeenth and eighteenth centuries, when the great European explorers, by opening the rest of the world to enquiring travellers, created the occasion for a second, similar programme of research and collection, under the direction of the Swedish naturalist Linnaeus.

Aristotle himself left descriptions of every aspect of organic life which can be studied with the unaided eye, and many of them were based without doubt on his own first-hand observation. He followed out, day by day, the slow changes within chick-eggs from laying to hatching; he was particularly knowledgeable about marine biology; and he gave a full account of the life of the honey-bee, referring in passing to the 'waggle-dance' performed by the bees on their return to the hive—the so-called 'language of the bees'. He built up an elaborate system for classifying the species of the animal kingdom, which began by dividing animals into two main groups: those having red blood, corresponding roughly to our own 'vertebrates', and those lacking it (the invertebrates). He subdivided these two groups again, according to the way in which the young are produced, whether alive, in eggs, as pupas, and so on.

The farther his biological experience extended, the stronger his fundamental convictions became. He was struck more and more by the tremendous variety to be found in the range of living forms: it was, he reflected, as though every single kind of living creature that *could* come into existence had in fact done so. It seemed easy enough at first sight to mark off certain broad classes of living things, all of whose main characteristics were markedly different from one another. But, as his researches continued, the lines between these main classes refused to stay clearly drawn: intermediate and ambiguous species kept coming to light, so that even the most obvious distinctions became blurred. To describe this continuity of organic forms, Aristotle used a figure of speech which was later taken up by mediaeval thinkers and put to very different uses. He spoke of the different organic species as lying along a Ladder or Scale of Nature:

> The transition by which Nature passes from lifeless things to animal life is so insensible that one can determine no exact line of demarcation, nor say for certain on which side any intermediate form should lie. As one goes up the ladder, next after lifeless things comes the class of vegetables, and these display quite different amounts of vitality; so that, in short, the whole vegetable kingdom, while lacking 'life' if compared with animals, possesses a great deal of 'life' by comparison with other bodies. Among plants then (as we said) a continuous scale of forms is observed leading up to the animal kingdom. In

> the sea, for instance, live some things of which one would be at a loss to say whether they were animals or vegetables. Some of these things are firmly rooted, and in several cases die if cut loose [e.g. sponges and sea-anemones] . . . [Likewise] throughout the entire ladder of animal forms one finds their vitality and ability to move about progressively increasing bit by bit; while the same is true about their habits of life. . . .
>
> Thus Nature passes from lifeless things to animals in an unbroken sequence, through a range of intermediate beings which are alive and yet do not represent true 'animals'; and so close together are all these neighbouring groups of creatures that there seems scarcely to be any clear distinction between them.

To a twentieth-century reader, Aristotle may here seem to be glimpsing our modern biological system, in which the natural sequence of forms on his Ladder of Nature reflects a temporal sequence—indeed, the evolutionary family-tree. The words 'Nature passes from lifeless things to animals . . .' might hint that, in earlier ages, Nature has passed step by step up the ladder of forms. Yet, for Aristotle's own purposes, there was nothing temporal—far less evolutionary—about this doctrine. The Ladder of Nature was never more than a striking fact about the *present* world of living things.

To Aristotle, the most significant things about organisms were the directedness of their activities, and the functional character of their structures. In order to bring out the importance of these features, he attacked the theory of natural selection put forward by Empedokles in the fifth century. According to this theory, the original organic population of the world comprised every conceivable combination of limbs and organs, generated spontaneously and at random, and the adaptedness of living things which have survived simply resulted from the elimination of ill-adapted forms. Aristotle quoted Empedokles' view:

> An objection presents itself: why should not Nature work, not for the sake of something or because it is better so, but just in the way that the sky rains—not in order to make the corn grow, but of necessity? What is drawn up must cool, and what has been cooled must become water and descend; and the result of this is that the corn grows. Likewise if a man's crop is spoiled on the threshing-floor, the rain did not fall *for the sake* of this—in order that the crop might be spoiled—but this result simply followed.
>
> Why, then, should it not be the same for the organs in nature: e.g. that our teeth should come up as they do *of necessity*—the front teeth sharp, fitted for tearing, the molars broad and useful for grinding down the food—and that they arose, not for this end, but merely as a

coincidental result; and so with all other organs in which we suppose that there is purpose? So, wherever all the organs turned out just as they would have done if they had come to be for an end, such things survived, being organized spontaneously in a fitting way; whereas those which grew otherwise, perished and continue to perish.

He stated this view, not—as Charles Darwin thought—because he supported it, but in order to *refute* it.

Yet in disposing of this theory, Aristotle was faced with a dilemma. If everything that happens by physical necessity must also be regarded as happening by chance—as in Empedokles' theory—how can one account for the fact that, generation after generation, living things regularly appear with such well-adapted organs? Might not some of the things that evidently happen 'for an end' also be the normal end-consequences of perfectly natural processes—happening, in that sense, 'of necessity'? That some things may happen *necessarily*, and yet be very far from *chance* events, was what Aristotle had to demonstrate.

His objections were all very well, so far as they went. But there was an element of cross-purposes in the argument. Empedokles had not been talking about the way in which, generation after generation, organs continued to develop in individual animals. Rather, he had been speculating about the first appearance of present-day species in a hitherto lifeless world. Within the framework of Aristotle's general theory, however, this question did not arise. Only individuals were born and died. Species were eternal.

So, once again, we see the Greeks being frustrated by the limitations of their time-scale. Darwin rightly saw in Empedokles hints of his own ideas; yet for all that, the two theories were radically different. Darwin explained existing species as the outcome of natural selection acting on a slowly-varying population over millions of years; but for Empedokles, as for Anaximander, discovering the origin of species meant explaining how life came on the Earth in the first place. He was, in short, describing the initial *beginning* of species, not (as Darwin did) their evolutionary *descent*. Natural selection thus operated, for Empedokles, on a random collection of organs that had come into existence by chance. This Aristotle found incredible—that the first generation of each species should have been formed by processes so completely out of harmony with those responsible for all subsequent generations. If the adaptedness of organisms made any sense, the processes of which they were the end-products must surely have been functional at every stage. Rather than accept so gross a discontinuity between the first and later generations, Aristotle preferred to believe that there had been no first generation at all—i.e. that the organic forms now existing represented a permanent feature of the natural order.

For more than two thousand years, this was the accepted view of reflective naturalists. Only the doctrine of fixed species appeared to be compatible with adaptedness. Indeed, until quite recently men were faced with a choice, not between Darwin and Aristotle, but between Empedokles and Aristotle, for the Darwinian option was not yet open. The modern theory of natural selection—as a process in which adaptedness and spontaneity collaborate to modify organic species *at every stage*—could be made plausible only against the vastly enlarged time-scale of nineteenth-century geology.

FURTHER READING AND REFERENCES

The character of classical Greek historical thought, and its relation to philosophy, are discussed at length in

J. T. Shotwell: *The History of History*
R. G. Collingwood: *The Idea of History* and *The Idea of Nature*

On the cosmological strand in Greek thought, see

J. Burnet: *Early Greek Philosophy*
W. K. C. Guthrie: *A History of Greek Philosophy*, vol. I

On the historical strand,

F. M. Cornford: *Thucydides Mythistoricus*

An outstanding discussion of early Ionian cosmology is

C. H. Kahn: *Anaximander and the Origins of Greek Cosmology*

For Plato's creation-story, see the version of the *Timaeus* in

F. M. Cornford: *Plato's Cosmology*

and also Sir Desmond Lee's translation in the Penguin classics.

Aristotle's attitude to the problem of creation is discussed in

W. Jaeger: *Aristotle*
E. Zeller: *Aristotle and the Early Peripatetics*

and his biological work in

D'Arcy W. Thompson: *Aristotle as a Biologist*
C. Singer: 'Greek Biology' in *Studies in the History and Method of Science*

Finally, on the Stoics and Epicureans, see

A. J. Festugière: *Epicurus and his Gods*

as well as the translations of Lucretius by R. C. Trevelyan, and in the Penguin classics.

3

The Authority of the Scriptures

During the last centuries before Christ, the Greek philosophers had begun to come to terms with traditional religious conceptions, but the full confrontation of philosophy and theology took place only after the beginning of the Christian era. In astronomy and chemistry, the scientists of Alexandria experienced little direct conflict with the theologians; in the 'history of nature', by contrast, the opposition of the new theology was immediate and complete. The reason is not hard to see. The *Book of Genesis* was one of the founding documents of Christianity, as it had been of Jewish religion earlier. All attempts to reconstruct the early history of the cosmos by rational enquiry—whether based primarily on arguments or observations—found themselves in competition with the Bible story, and appeared to call it in question. As an entrenched part of Christian doctrine, the Biblical account of Creation had to be defended against any rival account, so in the Alexandrian debates about the origin of the universe religious considerations predominated. Whatever was the case in physical and chemical theory, here at least a conflict between science and religion could not have been avoided.

It was in this field, too, that the hold of tradition on men's minds was to prove the most tenacious and long-lasting. From the fourteenth century onward mechanics, astronomy, chemistry and even physiology succeeded in establishing themselves in turn as fields for original speculation independent of theological supervision, and even appeared capable of lending important new weight to the Christian Revelation. In reconstructing the history of Nature, however, there was no real progress until the nineteenth century, and at once the enterprise collided head-on with Biblical preconceptions dear to many Christians. Yet the issues involved were not new. They had already taken shape in broad outline early in the Christian era, and this is where we must now turn our attention.

Christianity and History

The Jews believed themselves to be a people chosen by God, and so had always taken history seriously. For them, the historical process was a divine drama, in which they themselves had been cast in an important role. Their standing as a nation depended on the promises made by God to Abraham at a particular moment in their past history, and their hopes for the future were bound up with the fulfilment of this covenant. The resulting conviction, that the course of historical events had a profound significance, was the most important single legacy passed on by Judaism to Christianity. It has helped to shape the whole European tradition, contributing to that strain of moral earnestness and concern which is characteristic of so much European thought, even those systems which are nominally anti-religious, like Marxism-Leninism. So, from the very beginning, Christianity was a 'historical' religion, seeing the world as a stage for divine action, and the life of Christ as God's supreme intervention in its affairs.

Viewing the historical process as a unique stage for the divine drama, early Christians might well have considered the study of history as a pious duty—in the same way that seventeenth-century Protestant scientists were to consider the study of science a contemplation of God's plan as manifested in Nature. So the rise of Christianity might have given a powerful new impulse to historical thought. There was, however, a powerful counter-influence at work. Early Christianity took over from Judaism not only the Hebrew chronicles, but also a preoccupation with prophecy and apocalypse. For both Christians and Jews, the history of the world was not the slow unfolding of a continuous development, but a sequence of unique events, each of which broke abruptly with all that had gone before. Had the birth of Jesus represented the natural outcome of some phase of historical development, there would have been a motive for studying its antecedents: for the better these were understood, the better one could then have appreciated its significance. But the birth of Jesus was not (to use Aristotelian terms) the 'realization' of a historical 'potentiality': rather, it was the fulfilment of a supernatural prophecy. By it, the world had entered a brand-new phase, and all that need be known of earlier ages could be found in the sacred chronicles of the Old Testament. Accordingly, for all the importance which Christianity placed on certain key events in history, its first effect was to counteract the Greek sense of historical development and continuity.

Yet Christianity did retain one last link with the temporal process—one last investment in history. Although its preoccupation with eternity destroyed much of the motive for historical enquiries, the divorce of Christianity from the world of historical change was never absolute. For the possibilities of divine intervention were not exhausted. The birth of Christ might be the central event in history, but the Almighty was always free to intervene again in the affairs of His terrestrial world. Like those Jews who—having rejected Jesus' claims to be the Messiah—looked for a future Saviour to establish the Kingdom of the Righteous on Earth, many Christians also expected a Second Coming, by which Jesus would inaugurate the new Heaven and new Earth apparently promised in the *Book of Revelation*. But, once again, this New Jerusalem was something prophesied not predicted, a matter for faith rather than reason. The moment of its coming could not be foretold by rational reflection on the slow development of history: at best, it might give warning of its approach by omens, portents, or signs in the sky. Such an attitude to history was quite foreign to the Greek philosophical tradition. And, the more the Christian message came to depend on prophecies, portents and revelations, the more difficult it was for Christianity to coexist with the rationalistic elements in Greek science and philosophy.

Fundamentalism and Allegory

From the outset, then, Christians were divided in their attitudes towards Greek philosophy, with its sweeping claims about the powers of the human reason; and this division of opinion is particularly clear in their attitudes towards history. Though the Greek philosophers had made few positive discoveries about the past history of the cosmos, they had achieved a good deal in a *negative* way: they had shown, for instance, how difficult it was even to frame meaningful questions about the Beginning of All Things, and how slight was the rational foundation for all beliefs about early history. Christians, however, were committed to the creation-story of the Old Testament, and by Greek standards were therefore going beyond the limits of rational enquiry.

The theologians were, of course, fully aware of this fact, for their new religion was consciously based on revelation. Yet, even though one accepted the *Book of Genesis* as the Word of God, this still left one question open—whether this revelation was to be interpreted literally or allegorically. Over this point, theologians were divided from the beginning. Some repudiated the whole of Greek thought, ignored the philosophers' difficulties about the Creation, and interpreted the Bible story in the most literal-minded way possible. Others, feeling the force of the philosophers' arguments on their own minds, wished rather to make their peace with philosophy, even if that meant interpreting *Genesis* allegorically. The resulting tension between those who were prepared to compromise, and those who refused to do so, set a pattern which has been followed again and again, whenever new intellectual achievements have called for a re-interpretation of traditional Christian doctrine.

The attempt to reconcile *Genesis* and Greek philosophy had, in fact, started well before the birth of Christ, for the problem affected philosophically-minded Jews quite as severely as it did Christians. (Even the classical Greeks had been confronted by a corresponding problem, and Theagenes was reputedly 'de-mythologizing' Homer before 500 B.C.) Among Jewish philosophers, the task of adapting the national myths to the demands of reason culminated in the *Allegories of the Sacred Laws* by Philo of Alexandria, who was born about 20 B.C. In this book, Philo set out to draw back the veil of words concealing the allegorical truths of *Genesis*. On the surface, these words appeared to convey authoritative statements about historical events several thousand years ago: in fact (he argued) the first two chapters of *Genesis* were roundabout ways of expressing certain fundamental philosophical truths. As with the myths of Plato, their real aim was to describe the relations between the intellect and the senses:

> 'And the heaven and the earth and all their world was completed.' . . . Speaking symbolically, he [Moses] calls the mind 'heaven', since the natures which can be comprehended only by the intellect are in heaven. And sensation he calls 'earth', because sensation has a corporeal, and somewhat earthly constitution. Mental objects are called 'incorporeals', being perceptible only by the intellect. Those of sensation are 'corporeal', comprising everything perceptible by the external senses.

By reinterpreting *Genesis* as allegory, Philo could outflank Aristotle's objections to the idea of a Creation completely. Indeed, he was able to embrace these objections himself, using them as evidence that the passages under discussion *must* be intended allegorically:

> 'And on the sixth day God finished his work which he had made.' It would be a mark of great naïvety to think that the World was created either in six days, or indeed in Time at all; for Time is nothing but the sequence of days and nights, and these things are necessarily connected with the motion of the Sun above and below the Earth. But the Sun is a part of the heavens, so that Time must be recognized as something posterior to the World. So it would be correct to say not that the World was created in Time, but that Time owed its existence to the World. For it is the motion of the heavens that determines the nature of Time.

Having cleared Aristotle and the fundamentalists out of the way at a single stroke, Philo was able to offer his own alternative reading of the Hebrew creation-story:

> So, when Moses says, 'God completed his works on the sixth day,' we must understand that he is not speaking [literally] of a number of days, but that he takes six as a 'perfect number'. [Philo uses this term in the sense which it still retains in twentieth-century mathematics.] For it is the first number which is equal to the sum of its factors—the half (3), the third (2) and sixth parts (1)—and is produced by multiplying the two unequal factors, 2 and 3. [i.e. $1+2+3=1\times2\times3=6$.]

Though none of the Early Fathers went quite so far as Philo in reinterpreting sacred history, one of them at least followed him some way along that path. On Origen, an Alexandrian Greek of the third century, fell the task of defending Christianity against powerfully-argued attacks from Porphyry and Celsus. Celsus had criticized the Christian Scriptures as bad history: how could one take on trust legends supported by so little evidence as those in the Bible? Origen conceded in reply that the testimony of the Scriptures was largely uncorroborated, but counter-attacked by arguing, with some justification, that the same objection applied to a great many historical traditions:

> Suppose someone were to assert that there never was any Trojan War, because of the impossible story interwoven [with the *Iliad* account of the War] about a certain Achilles being the son of a sea-goddess Thetis and a human Peleus . . . Bearing in mind the weight of fictions which have become attached to it (I know not how), how should we prove that it had in fact occurred, and that—as everyone believes—there was really a war at Ilium between Greeks and Trojans?

All one could reasonably do, he concluded, was to accept the main outline of Biblical history, as it had come down to us, reserving the right to explain away as corruptions or accretions any details which by now appear frankly incredible. In particular, one was not obliged to treat the Creation-story in the first chapter of *Genesis* as a literal account of an historical sequence of events.

The New Chronology

The radicals, however, represented only one wing of Christian thought. In its early days, Christianity struck firm roots more among humble, literal-minded men than among intellectuals, and sophisticated styles of theology never appealed strongly to these simpler believers. They saw no objection against accepting the Old Testament as a straightforward, authoritative chronicle of the world's history since the first day, and by about A.D. 300 the historical framework of the Christian world-drama had been fixed in the form which it was to retain for some fifteen hundred years. At the Council of Nicaea, called by the Emperor Constantine in A.D. 325 to define the official doctrines of the newly-established religion, the man who presided over the inaugural session was Eusebius. He was both a statesman and a sacred historian and, for all his intellectual commitments as a Christian, he knew very well how to handle and criticize historical sources. His *Chronology* and his *Church History* were both supplied with a formidable array of scholarly arguments and references.

The *Chronology* was a comprehensive world-history, collecting together the key-documents recording the histories of the different ancient civilizations, and setting out their dating-systems in a comparative time-chart. (In this way, Eusebius established an intellectual method which was to be followed many times over: Isaac Newton himself employed the same pattern nearly fourteen centuries later in his own *Chronology of the Ancient Kingdoms Amended.*) The first part of Eusebius' book, comprising the source-material, soon disappeared from sight and has become available again only during the twentieth century, through the rediscovery of an Armenian translation. The chronological tables accompanying it, on the other hand, were passed down from generation to generation throughout the Middle Ages. They well deserved to be, for they represented a more detailed, critical and painstaking exercise in comparative history than any that had gone before. Eusebius argued that the traditional Jewish chronicles were quite as credible and trustworthy as any of their rivals: given the necessity of choice, the Christian was entitled to accept them as the foundation of his own chronological scheme. Eusebius himself began his chronicle with Abraham, diplomatically evading the more serious intellectual problems raised by *Genesis*. The omission was made good by his Latin translator, Jerome, who counted 2242 years from Adam to the Flood, and 942 from the Flood to Abraham—though later he revised these figures to 1656 and 292. With Jerome's additions, the chronology of Eusebius provided the numerical time-scale on which, from then on, historians in the Western world founded their dating-system.

But, as so often happens, the unsophisticated found their inspiration elsewhere. The simplified version of cosmic history which was popular throughout the Middle Ages went back before Eusebius, to the *Chronographia* of Julius Africanus. Julius took from Jewish literature the idea of a 'millennium'—the thousand-year-long Kingdom of the Messiah which prophecy declared would end the history of the world—and used it as a symbolic key for interpreting Old Testament chronology. On this interpretation, the whole of history corresponded to a cosmic week, each of whose days lasted a thousand years. (The justification for this step was found in the words of Psalm 90, verse 4, 'A thousand years in thy sight are but as yesterday'.) The first six days of the present order covered a period beginning with the Creation and lasting up to the Second Coming of Christ, which would inaugurate the final day, or millennium. To fix the time-scheme precisely, one had only to decide how many cosmic days had elapsed between the Creation and the birth of Christ. Julius assumed that this period had lasted five-and-a-half days (5500 years), which implied that the Second Coming could be expected around A.D. 500. Although the year 500 came and went without any sign of a millennium, this 6000-year time-scheme survived with modifications right up to the Reformation and was taken for granted by most devout Protestants, including Luther himself.

Yet any literal belief in an *earthly* millennium was never fully orthodox. This is clear from Augustine's great treatise of Christian doctrine, *The City of God*, begun in A.D. 415. Augustine's thought operated on two levels. He contrasted absolutely the present order of temporal things on Earth with the eternity of God's nature in Heaven. As to the Earth's history, he recognized only the authority of Holy Scripture, and so rejected the pagan chronicles wherever they contradicted it; yet he also denied that the New Jerusalem referred to in *Revelations* represented the promise of an earthly Paradise. To treat the Second Coming as an historical event still in the future, which would inaugurate an era of peace, plenty and pleasure on Earth, was far too materialistic a view. On the contrary, the *Book of Revelation* was an allegory of the true 'world to come'—in Heaven. So one should avoid taking prophetical allusions to a 'millennium' literally, and reinterpret the references to 'a thousand years' in *Revelations*, in the same way that Philo had done with the *Genesis* story of the 'six days' of Creation:

> Now, if 'a hundred' be sometimes used to designate perfection . . . why, then, may not 'one thousand' represent consummation—the more so, seeing that it is the most solid square figure that can be constructed from ten? [i.e. Ten cubed equals one thousand.] So that we interpret that passage in the Psalm, 'He hath always remembered His covenant and promise that He made to a thousand generations', by taking 'a thousand' to mean 'all in general'.

From now on, the task of the Christian scholar was to expound these Biblical allegories, and make their rich and complex message plain to the unlearned. In this exegesis, nothing could be assumed to mean only what it seemed on the surface to mean. Animals, plants, planets, precious stones, geometrical figures, numerical ratios: every allusion had a hidden significance which the serious student of the Bible must struggle to demonstrate, and any instrument which could assist this task was justified—even the previously despised learning of the pagans. It was in this way that the science and philosophy of Greece, Mesopotamia and Egypt at last earned a place in the body of Christian learning. It was of course a subordinate, even subservient, place. Until A.D. 1000, philosophy remained quite unambiguously the handmaiden of theology, and natural science had little to do except fetch and carry for the Biblical interpreters.

All the same, this new tolerance of science and philosophy was a worth-while concession. As all knowledge of Nature might now be turned to worthy purposes, an interest in astronomy or zoology was no longer a stumbling-block between the Christian and his salvation. But it did not take would-be scientists very far, since observations were still accepted as interesting only for the light they threw on Church doctrine. Mediaeval bestiaries, as a result, jumbled together natural history and theology:

> The Ant has three peculiarities. The first is that these creatures walk in a line, and each of them carries one grain in his mouth. Their comrades do not say 'give us of your grains' to the loaded ones, but they go along the tracks of the latter to the place where they found the corn, and they carry back their own grain to the nest.
>
> Mere words, you see, are not an indication of being provident. Provident people, like ants, betake themselves to that place where they will get their future reward.
>
> The second peculiarity is that when an ant stores seeds in the nest it divides them into two, lest by chance they should be soaked with rain in winter, and the seeds should germinate, and the ant die of hunger.
>
> Oh Man, divide you also the words of the Bible in this way, i.e. discern between the spiritual and carnal meanings, lest the Letter of the Law should be the death of you. It is as the apostle observes: 'for the Letter kills, but the Spirit gives life.' The Jews, attending only to the letter, and scorning the spiritual meaning, have been killed with hunger.
>
> The third peculiarity is that in time of harvest an ant walks about among the crops and feels with its mouth whether the stem is one of barley or one of wheat. If it should be barley, it goes off to another stem and investigates; and if this feels as if it were wheat, it climbs up to the top of the stem and, taking thence the grain, carries it to its habitation. For barley is the food for bigger beasts.
>
> This was why Job said: 'Instead of wheat, it produced barley to men'—that is to say, the doctrine of heretics did. For those things which shatter and kill the souls of men are like barley, and meet to be thrown far away. Fly, O Christian, from all heretics, whose dogmas are false and inimical to the truth.

Preoccupied as they were with allegory, Augustine's successors did little to resolve outstanding philosophical difficulties about history and cosmology: rather, they side-stepped them. During these centuries, indeed, many of the works and even the very names of the Greek philosophers came to be forgotten. Plato's *Timaeus* was preserved, for the sake of apparent anticipations of Christian doctrine. (Augustine discussed, in all seriousness, whether Plato might have known the Old Testament.) But much of Aristotle was lost or remained unread, until after A.D. 1100. This neglect was not primarily due to malice, for by A.D. 500, the worst of the anti-Greek riots were over. It was simply that Christians no longer had any use for the old questions. Given the divine authority of the Biblical narrative, most European scholars were content to accept its time-span of six thousand years at its face value. Why should they do anything else? They had nothing to put in its place.

In one part of the globe, however, an interest in scientific and historical questions for their own sake meanwhile remained alive. At Baghdad and elsewhere, Islamic scholars—notably al-Kindi, al-Farabi and Avicenna—elaborated on the classical Greek discussions about time and eternity, the origin and structure of the heavens, and the formation of the Earth, and in so doing they immediately faced intellectual problems which were to arise for Christian theologians only after A.D. 1100. The problems of time, for instance, were just as difficult for Muslims as they were to be for mediaeval Christians, committed as they both were to a belief that the world is God's creation. Before Aristotle's philosophy could be assimilated into Islam, some way had to be found of circumventing his conclusion that the material world could not have come into existence at any one moment in time. The task was not easy. Al-Kindi denied that past time could have been infinite, but could think of no reason other than a Divine act of will why it should have begun just when it did: God was surely free to create the world whenever He chose.

Avicenna rejected this suggestion on the grounds that it was far too arbitrary. God must always have some reason to act as He does, and no particular moment in the characterless history of a primal void could be more or less 'suitable' for the Creation than any other. God's apparent choice must be an intellectual illusion, whose sources could be appreciated if one analysed more carefully what is meant by calling God the 'creator'. At this point, Avicenna was moving in a direction which Thomas Aquinas, the great Christian philosopher, was to follow three hundred years later. The fundamental error was to think of God creating the world 'from outside', like a potter shaping a pot. Christians, or Muslims, are not compelled to interpret the Divine Creation in this naive way. To speak of God 'creating' the world was certainly more than an allegory but, when compared with human ideas of creation, it was at most an *analogy*. No doubt the material world did have a temporal beginning, but this was not the crucial thing: rather, the significance of the Creation lay in the world's continuing dependence on God. Apart from God, the world could not go on existing. So, even the most rational arguments of the Greek philosophers were gradually integrated into an all-embracing system of theological symbols and analogies.

Yet theological motives were never completely dominant among the Islamic philosophers—at any rate before the twelfth century. In little more than a hundred years the Muslim conquests had, with scarcely a check, carried Islam as far as the Atlantic in one direction, and to the borders of India in the other. The very rapidity of this expansion created an atmosphere of confidence, which extended to the intellectual field also. In the great period of Islamic science and philosophy one finds little of that defensive conservatism with which Augustine reacted to the sack of Rome. Among the unchallenged splendours of Baghdad and Samarkand, Muslim rulers patronized science and scholarship in a liberal spirit, and Islamic scholars let their curiosity and imagination roam with comparative freedom. All good Muslims were, of course, concerned with the proper interpretation of the Koran, but other fields of learning were not subservient to theology. As a result, Arabic doctors, astronomers and chemists built up novel and enduring structures of their own on the intellectual foundations laid by the Greeks. Even in the history of Nature, Islamic scientists were beginning to see farther than their predecessors had ever done. Around A.D. 1000 Avicenna was already suggesting a hypothesis about the origin of mountain-ranges which, in the Christian world, would still have been considered quite radical eight hundred years later:

> Mountains may be due to two different causes. Either they are effects of upheavals of the crust of the earth, such as might occur during a violent earthquake, or they are the effect of water, which, cutting for itself a new route, has denuded the valleys, the strata being of different kinds, some soft, some hard. The winds and water disintegrate the one, but leave the other intact. Most of the eminences of the earth have had this latter origin. *It would require a long period of time for all such changes to be accomplished, during which the mountains themselves might be somewhat diminished in size.* But that water has been the main cause of these effects is proved by the existence of fossil remains of aquatic and other animals on many mountains.

For the time being, these insights were not followed up, and the full antiquity of the world remained unsuspected.

The Mediaeval World-Allegory

From the eleventh century on, it was the turn of the Christian West to gain in confidence and in intellectual freedom. The city of Constantinople, which had been the political capital of the Eastern Roman Empire since the fourth century A.D., was still far larger and richer than London or Paris, and even than Rome itself; but it was beginning to feel the strain of the recurrent wars against the Slavs to the north and the Arabs to the east. Soon the Arabs in their turn, having been checked in Europe, were taken in the rear by the Mongols: the sack of Baghdad in 1258, accompanied by the destruction of the ancient Mesopotamian irrigation-system, delivered a blow to Middle Eastern civilization from which it has not fully recovered even now. Behind the Mongols came the Turks, who before long took over all the Arab conquests up to the borders of Morocco, and destroyed the last political power of Constantine's Byzantine Empire. In the countries of Western Europe, by contrast, the Islamic threat had been first held, then slowly pushed back. From being poverty-stricken provinces of a Roman Empire overrun by the barbarians, the countries of the West became powers in their own right, with a steadily-growing economic strength and political influence; and a fresh community of scholars grew up, centred on the newly-founded universities at Paris, Bologna and Oxford.

In the minds of these men, a new vision of the natural world took form, which we know as 'the mediaeval world-picture'. It was a vision that allowed scientific and philosophical issues a more significant place than Augustine had ever done. By the middle of the thirteenth century, indeed, Thomas Aquinas and Albert the Great could set about turning Aristotle into a good Christian, with the same determination that al-Kindi and Avicenna had devoted to reconciling his philosophy with Islam. Yet the mediaeval vision of Nature never wholly lost the symbolical character of earlier Christian thought—never entirely ceased to be allegory. Though, for want of better information, the picture it gave was taken as being true in point of fact, this literal truth was not its prime value or significance: one learns more about the mediaeval world-picture from a poem like Dante's *Divine Comedy* than from any single prose treatise of the period. For, although the facts of Nature were at last acquiring a value in their own right, they were still valued more highly as religious symbols. The practice of looking for hidden spiritual meanings in the simplest and humblest occurrences was deeply engrained, and it turned the whole world of Nature into a subject for allegorical interpretation.

Mediaeval thinkers abandoned the Early Fathers' attempt to entrench the flat-earth theory into Christian orthodoxy. Instead, they took over from the ancient astronomers the theory that the Heavens comprised some sixty concentric translucent spheres, the Earth being at the centre and the sphere of fixed stars enclosing the whole system, with a diameter of about a hundred million miles. Yet this picture did not win their allegiance on account of its scientific merits alone—as being the hypothesis least at variance with the facts of observation. (Scientifically speaking, the intellectual difficulties left unresolved by Ptolemy continued to embarrass professional astronomers right up to the time of Copernicus.) Quite as important, and frequently more so, was the way in which the spatial layout of this astronomical picture symbolized the degrees of perfection of 'earthly' and 'heavenly' things. Christians of course rejected the older Middle Eastern worship of planet-gods, yet the poetic values of the astral religions carried over into Christianity in part, and they were reinforced by Aristotle's theory of the 'quintessence'—the imperishable substance of the heavens, exempt from the change and decay afflicting terrestrial things. By Dante's time, the astronomical 'nest' of crystalline spheres had become the accepted symbol for the scale of perfection. Things on Earth were 'lowest', both in location and in worth. The passage 'upwards', through the spheres of the Moon, Sun and successive planets to the outermost heavens, represented a journey from corruption to perfection. Outer space was a far, far better place than anywhere on Earth, and the height of blessedness was to be 'transported' as far as 'the seventh heaven'. (This mediaeval scale of values has left an enduring mark on the metaphors and idioms of twentieth-century speech.)

Aristotle's zoology was reinterpreted in a similar allegorical way. He had recognized the difficulty of marking off different living species into clear, distinct groups, and had concluded from their unbroken sequence that the individual steps in the Ladder of Nature were insensibly small. This observation was generalized by his successors into the so-called 'Principle of Plenitude', according to which the sequence of natural forms was not just sensibly continuous, but *absolutely* so. From the humblest and most formless creatures at one end, up to the highest and most complex at the other, there extended a single scale of beings. But, once again, what had begun as scientific relationships were found to conceal—or reveal—deeper spiritual truths. Aristotle's Ladder of Nature turned into a hierarchy, the continuous scale of beings into a 'chain of command'. The more complex creatures were 'higher', not just in a bald zoological sense, but because, in the divine government of the world, they were the 'superiors' of the 'lower' creatures—set in authority, and having dominion over them. Human beings had a 'divine right' to domesticate or slaughter the brute beasts; and the higher animals in turn were entitled to subjugate the lower ones; and so on down through the scale. Moreover, the ladder itself now extended upwards in a way Aristotle certainly never envisaged, passing beyond Man, through a whole range of different spiritual beings, and up to the Deity at the very head. At this point, of course, zoology was left far behind. The whole point of the doctrine now lay in its ethical and theological content. The Almighty was not only the Creator of the world, but also its Commander-in-Chief, and Man's place in the divine hierarchy subordinated him to God and the angels, as surely as it set him in authority over the lower Creation.

These two elements in the mediaeval world-picture dovetailed neatly together. The scale of perfection symbolized in the geocentric astronomical scheme fitted happily with the religious hierarchy based on the Ladder of Nature. Plants were rooted into the lifeless earth, worms and snakes lived on its surface, quadrupeds were lifted slightly above it, while Man's erect posture symbolized his unique spiritual ambition—to raise himself upwards towards higher, heavenly things. The space between the sphere of the Moon and the outer heavens was filled with the upper hierarchy of spiritual beings—the immaterial, invisible powers, angels and archangels. Before long, indeed, these became identified with the 'intelligences' which al-Farabi had considered the motive agencies of the planetary spheres.

Other ideas inherited from Alexandria fitted naturally into the same picture, like pieces in a jig-saw. Philosophers in late antiquity had made the familiar interaction between the Heavens and the Earth the basis for a complete system of correspondences; each of the heavenly bodies was associated with particular plants, animals, precious stones, minerals, parts of the human body and so on. The interconnections which Stoic philosophers had treated as links in a deterministic network were now transformed into channels for the divine power. Like some great source of invisible energy, the Deity directed His influence from the outermost Heavens right down to the Earth, along the 'conducting paths' established by these correspondences. Everything in the greater world of the Heavens (= *macro-cosm*) had its counterparts in the lesser world on the Earth (= *micro-cosm*). The different metals, the different plants, the limbs and organs making up the human frame all operated under the aegis of, and in harmony with, their symbolic counterparts in the Heavens. It needed only one short step, and the whole of astrology and alchemy could have been incorporated into the orthodox mediaeval picture of Nature. In fact, this step was not taken, and throughout the Middle Ages these activities remained under the suspicion of heresy. Yet the objections to astrology and alchemy had little to do with their *concepts*. Those same beliefs about the interaction between the celestial and terrestrial worlds which had been their starting-points were important for orthodox mediaeval thought also. Rather, it was the *ambitions* of the two sciences that kept them under a cloud. Both were suspect, as being excessively presumptuous. The astrologer was seeking to predict things that God alone could foresee, and the alchemist aimed at power over Nature of a kind reserved to the Deity.

To pursue the mediaeval allegory still farther: the Stoics in the ancient world regarded the material substance of the Heavens as more 'spiritual' and 'ethereal' than the substance of the Earth and terrestrial things. This contrast too was swept up into the mediaeval imagery. In antiquity, of course, the Stoic doctrine had directly contradicted Aristotle's absolute distinction between the quintessence of the Heavens and the four terrestrial elements: for the Stoics, the difference between the material of the Heavens and the Earth was only a matter of degree. But this flat contradiction did not prevent the allegorists from finding spiritual truths in both doctrines. The purer and more spiritual substance of each successive heavenly sphere reflected a more refined and potent 'intelligence', and this sequence once again culminated in the perfection of the Divine Mind.

In this way, pieces from the whole corpus of ancient science and philosophy were brought together and combined in a single symbolic picture of the world. Seen against the background of this allegorical world-picture, the intellectual preoccupations of mediaeval scholars are understandable. For their central question concerned the relationship between the Divine Mind and the minds of individual humans. The world was an organism whose continued existence was sustained by the Divine Act, and it seemed essential to measure the value and authenticity of human knowledge by tracing it back to the presumed Source of all intelligence. This was the starting-point for a great deal of metaphysical debate, but it acted also as an incidental stimulus to physics. Many of the scholars involved in the discussion recognized that the sense of sight was the pre-eminent means by which we acquire knowledge, and identified *light* as the spiritual bond linking the individual human intellect to its divine origin.

Every aspect of light had its symbolic associations, and optics, as a result, was the branch of physics in which mediaeval scholars did most original experiments. Again and again they were drawn to the problem of the rainbow—a phenomenon which had had symbolic significance in all ages. Spiritual power, likewise, was thought to confer a special 'radiance' on its possessor, which he could transmit to others by 'the light of his countenance', so that at times it seemed to glow around his head, in a halo of radiation. The Sun, illuminating our minds as well as the Earth, retained for mediaeval Christians much of the significance it had possessed in earlier mythological times. It might no longer be a god, but at any rate it was the source of Divine Light. So Francis of Assisi, though no metaphysician, could acclaim the Sun in his *Canticle of All Created Things* as the first of God's creatures:

Laudato sie, mi Signore, cun tutte le tue creature
spetialmente messor lo frate sole
lo qual jorna et allumini noi per loi.
Et ellu è bellu e radiante cun grande splendore
de te, altissimo, porta significatione.

Be praised, my Lord, with all your creatures,
especially master brother sun, who brings
day, and you give us light by him.
And he is fair and radiant with a great shining—
he draws his meaning, most high, from you.

Such, then, was the natural world as pictured by the scholars of mediaeval Europe. The main purpose of the picture was to serve as an allegory of the divine Creation, rather than as a painstaking deduction from scientific observation, but it was also accepted incidentally as being a literal description of Nature. Yet, allegory or no, the mediaevals were forced, like their Islamic predecessors, to face the problems that arose when Aristotle's science was turned to theological ends. That the world was created by God was an axiom. Yet, as Avicenna had known, this axiom could not easily be reconciled with Aristotle's teachings about Time, and the theologians of mediaeval Christianity were hard pressed to match an orthodox view of Creation in with the rest of Aristotle's science.

The compromise eventually established by Thomas Aquinas depended on making the doctrine of Creation a matter for revelation rather than reason. Aristotle had not really proved that a beginning of the universe in time was an absolute impossibility: rather, he had demonstrated that reason alone could not *prove* that such a beginning had occurred. If this were so, reason alone could not establish, either, that it had *not* occurred. The question was not one for logicians to argue out rationally, but had to be decided by other means—and Christians were fortunate, in having the answers revealed to them in *Genesis*.

This compromise was achieved only at a price. Questions of revelation had to be taken on trust, and to that extent they ceased to be open to scientific or philosophical discussion. The full consequences of this fact have not always been appreciated. Aquinas himself, for instance, set out five arguments (or *viae*) by which a man could come to understand the nature of God, and one of these involved the idea of Creation. This 'cosmological argument' declared that the world, being God's Creation, was absolutely dependent on Him for its existence. Now just how much was entailed by this statement? At first glance, it might mean: 'The Christian conception of God entails that He *brought* the world *into* existence at some time.' In fact, the same distinction has to be made here as in the case of the *Timaeus* and Avicenna. All the cosmological argument entails is a *continuing* dependence of the world on God: by itself, it leaves all questions about the temporal origin of the world entirely open. For the orthodox Christian, the fact that the world existed at all was sufficient proof of the continuing action of God; but one could establish whether it came into existence at some definite time, or had continued from eternity, only by appeal to divine revelation in Holy Scripture.

Serious theologians have always taken care to respect this distinction, but in popular handbooks the step from 'God the ground of all being' to 'God the initial maker of all things' is sometimes glossed over. Yet this step is of great significance for our present enquiry. For, philosophically speaking, the world-picture of mediaeval Christendom was scarcely more historical than that of Plato's *Timaeus*. All that could be known about the remote past of the cosmos was contained in the sacred chronicles of the Bible, and men in A.D. 1300 were as far as ever from reasoning their way past the historical time-barrier. Different scholars reacted to the position with different attitudes. Some, like Albert the Great and Vincent of Beauvais, continued to speculate and theorize about the problems of zoology and geology. Others, like Nicholas of Cusa, found a higher wisdom in recognizing the limitations of the human reason. Man was powerless to rise from a study of the finite things of the world to the infinitude of God, and all attempts to build up a scientific cosmology must fail. God was the 'meeting-point of opposites', at once the 'centre' and the 'circumference' of the world, about whose overall form men could prove nothing, but only speak symbolically.

The Dream of the Millennium

Meanwhile, orthodox theology was not having everything its own way. The arguments of the learned doctors were too sophisticated and their conclusions too negative to satisfy popular appetites. Alongside the official Catholic doctrines, as formulated and clarified by Augustine and Aquinas, a more dramatic and prophetic interpretation of the Christian message continued to flourish—though often clandestinely. Jesus' words about the Coming Kingdom of God had always raised expectations of an imminent transformation in the present order of things. (As revivalist placards still say, 'The End of the World is at Hand'.) For a long time, people believed that the promised Kingdom of the Saints would be established in their own lifetimes; and in periods of persecution this confidence—or desperate hope—spread far and wide, encouraging Christians to brace themselves for the coming struggle, when the faithful would be rewarded and their persecutors cast down.

Two factors told against the belief in an imminent apocalypse. The very success of the Church, in gaining an established position in the Roman world, had made revolutionary ideas unpopular with the ecclesiastical authorities; and the simple passage of time had done much to reduce the dramatic tension of the first centuries A.D., lengthening the temporal perspective in which Christians looked at the history of the world. Yet the dream of a New Era of absolute justice, to last a thousand years, was not so much stamped out as driven underground, and survived as a continuing influence on the popular imagination. In Germany, Bohemia and the Low Countries, its hold was particularly tenacious. Again and again throughout the Middle Ages, wandering preachers were able to win a following by prophesying the imminent downfall of the established political order, and the beginning of a new reign of justice for the faithful and downtrodden.

Much of the popular enthusiasm for the Crusades sprang from this source. The thousands who marched eastwards towards the Holy Land were not only trained and disciplined contingents, led by the noblemen of Europe. They included also unorganized hordes of poor landless peasants, unemployed townsfolk and even children, for whom the struggle against the infidel was a preparation for the return of Jesus to a New Jerusalem. The ragged crowds following in the steps of men like Peter the Hermit were an embarrassment just as much to their own nominal rulers as they were to the Byzantine authorities at Constantinople. Lacking all discipline and order, living off the land as they went, they appeared to outsiders as an obnoxious rabble, to be slaughtered or moved on as quickly as possible. But in their own eyes they were the servants of the Lord—

> Singing songs of exaltation,
> Marching to the Promised Land.

A few of them reached Palestine, but many perished on the way. Some who were frustrated in their ambition to recover Jerusalem from the infidel were persuaded to turn their anger on to the Jews of their own countries. Thus was established that pattern of proletarian discontent, fanaticism, anti-Semitism and belief in a 'new order' which, lasting on right into the twentieth century, was exploited by the revived beastliness of Hitler's National Socialism.

The threat to Church and State posed by these popular movements was never an empty one. To the underdogs of mediaeval Europe, the prospect of the Second Coming had the same appeal that the 'withering-away of the state' had for primitive Marxists. Just as the ideological history of Communism cannot be wholly divorced from the political history of its times, so also mediaeval philosophy and theology must be viewed against a similar counterpoint of political stress and relaxation. When, around A.D. 1200, the lost works of Aristotle began to circulate once again, they were at first suspect not merely on account of their novelty, but even more because their reappearance unhappily coincided with a debate about the 'Amaurian' heresy—named after Amaury of Bène, a scholar at the University of Paris. A synod of the Church investigating the heresy suggested that the newly-published translations, with their commentaries by Averroes, had encouraged this deviation, and for much of the thirteenth century their use was forbidden.

Frequently, prophecies of the millennium were supported by numerical arguments recalling those of Julius Africanus. Towards the end of the twelfth century, Joachim of Fiore calculated that the Second Coming would be due sometime in the years between 1200 and 1260: since forty-two generations had passed between Abraham and Christ, a new dispensation might be expected a further forty-two generations after the birth of Jesus. For a time, Messianic hopes were concentrated on the Emperor Frederick II, the young grandson of Frederick Barbarossa, who had died in 1190 on the Third Crusade. Both Fredericks in fact became legendary figures. For centuries, it was rumoured in parts of Germany that Barbarossa himself had never really died, but was in hiding, and that he would return in majesty to put down corrupt priests and defend the poor against the rich: these rumours inflamed popular imagination especially at times of misery and distress—for example, during the Black Death of 1348–9. Sometimes the prophecies forecast the date of the coming millennium with great precision. In Bohemia, for instance, many people expected the towns and villages of the unfaithful to be destroyed by fire during the five days beginning 10th February 1420. Though never fulfilled, these Messianic hopes were never extinguished, merely deferred. In the sixteenth century, such dreams were still widespread, and the success of the Lutheran Reformation gave them an encouragement which Luther himself deplored.

Thus the revival of scientific curiosity and secular learning in Western Europe found Augustine's teachings about the City of God still far from established. The expectations which the Council of Ephesus had attempted to stamp out were still alive, especially in those parts of Europe that had now broken away from the Catholic Church; and the chronological arithmetic on which these ideas fed remained a powerful influence on seventeenth-century thought. For even Augustine had not foreseen the present order of things enduring for more than a thousand years, and the mediaeval time-scale remained nearly as short as the one worked out by Julius Africanus. This time-scale was very difficult to change. All conjectures based on human observation appeared very flimsy, when compared with the categorical statements of Holy Scripture. So long as human reason weighed no more than thistledown when put in the scale against divine authority, little could be done to extend the scope of the world-picture. In the chapters which follow, we shall see how this traditional view was eventually weighed against the testimony of Nature itself; and how, as a result, the narrow limits within which the whole of cosmic history had up to then been confined were first eroded, and finally swept away.

FURTHER READING AND REFERENCES

For the first Christian ideas about history, and their relation to earlier religions, see

The Idea of History in the Ancient Near East: ed. R. C. Dentan
F. Cumont: *The Oriental Religions in Roman Paganism*
J. T. Shotwell: *The History of History*

On the general relevance of history to Christian belief and *vice versa*, see

H. Butterfield: *Christianity and History*

Augustine's *City of God* is available in numerous editions: the most relevant sections are to be found in Books XI, XII, XVIII, XX. There is a recent discussion on the Arabic philosophers' attitudes to the creation in

S. M. Afnan: *Avicenna*

The development of mediaeval Europe, as a background to mediaeval thought, is discussed in

R. W. Southern: *The Making of the Middle Ages*
H. Pirenne: *Economic and Social History of the Middle Ages*

The intellectual development is covered in

David Knowles: *The Evolution of Medieval Thought*
G. G. Coulton: *Studies in Medieval Thought*
C. H. Haskins: *Studies in the History of Medieval Science*
H. O. Taylor: *The Medieval Mind*
Charles Singer: *From Magic to Science*

For different aspects of the 'world-allegory', consult

A. O. Lovejoy: *The Great Chain of Being*
T. H. White: *The Medieval Bestiary*
E. M. W. Tillyard: *The Elizabethan World-Picture*
A. C. Crombie: *Robert Grosseteste* (on the metaphysics of light)

The standard analysis of the millenarian movements of the Middle Ages is the excellent book

Norman Cohn: *The Pursuit of the Millennium.*

4

The Revival of Natural Philosophy

HISTORIANS traditionally mark the end of the Middle Ages by one of two events: the capture of Constantinople by the Turks in A.D 1453, and the proclamation of the Protestant Reformation by Martin Luther in A.D. 1517. But, inevitably, the exact placing of such historical milestones is arbitrary. The renaissance of learning was not a sudden or abrupt event, but a continuing process some of whose origins go back as far as the twelfth century. So we should not be misled by the popular picture of a mediaeval culture absolutely dominated by theology, or suppose that, in the sixteenth century, it was suddenly replaced by a new system of progressive thought unhampered by out-of-date authorities. This account of the renaissance was always a polemical caricature, originated and put into circulation around A.D. 1600 by Francis Bacon and his fellow-prophets of secular learning.

Still, the sixteenth and seventeenth centuries did see a tremendous acceleration in the processes of intellectual change, following the adoption of the printing-press into Europe. (It had been in use in China for several centuries.) The spread of leisure and learning among the lay public opened up fresh markets for the classics of antiquity, and the laborious method of copying manuscripts could not keep pace with the new demand: in return, the introduction of the printed book stimulated intellectual activity among people who would scarcely have been literate two centuries earlier. Thus secular learning and printing reinforced one another, forming the twin foundation-stones of a new era, and re-shaping European culture even more radically than their twentieth-century counterparts of universal education and the 'mass media'.

Yet, to begin with, this transformation of the intellectual situation contributed little to the history of Nature. Far from taking men directly to our own dynamic vision, the first currents of fresh thought flowed in a contrary direction. By 1730, many scientists of Western Europe had come to accept a view of Nature even more static and fixed than that of mediaeval Europe, and the traditional six-thousand-year time-scale still retained much of its earlier authority. The views that Darwin in due course refuted were thus not the superstitious relics of mediaeval religion, but a rival scientific theory built up, during the seventeenth and eighteenth centuries, by men who prided themselves on their own modernity.

The Fall of the World

The conception of a fixed Order of Nature, which played so large a part in science in the years around 1700, was the product of two factors, each of which tended to strengthen the other. One of these was the continued acceptance of the Biblical time-scale: the other was the adoption of mathematics as the foundation of physical thought, following the example of Galileo and Descartes.

The first outcome of the Protestant Reformation had in fact been to establish the Biblical time-scale even more firmly than before. Throughout the high Middle Ages, the literal text of the Bible was only one of two unquestioned authorities, both of which were equally important. The second was the tradition of the Church, and a thousand years of Catholic teaching and allegorical interpretation had weakened any temptation to treat the words of Scripture in too literal-minded a way. But, in the eyes of the Protestant reformers, this Church tradition was thoroughly corrupted, and their ambition was to revive an older and supposedly purer Christian tradition. As a result, they were thrown back on to the words of the Bible itself far more forcibly than Catholics had ever been. So Luther and his successors began to treat the early books of the Bible as an authoritative historical record to an extent scarcely contemplated in earlier centuries.

This new fundamentalism treated *Genesis* as the supreme textbook of cosmology and geology. From now on, the Biblical Flood was regarded as the most powerful single agent responsible for geological change. Luther put it as follows:

> As therefore since the Flood mountains exist where fields and fruitful plains before flourished, so there can be no doubt that fountains and sources of rivers are now found where none existed before, and where the state of nature had been quite contrary. For the whole face of Nature was changed by that mighty convulsion.

The Deluge had destroyed the Garden of Eden along with everything else, and the corruption which began with the Fall of Man had spread to the inanimate world:

> Even the Earth, which is innocent in itself, and committed no sin, is nevertheless compelled to bear sin's curse. . . . All creatures, yea, even the Sun and the Moon, have as it were put on sackcloth. They

> were all originally 'good', but by sin and the curse they have become defiled and noxious.

By A.D. 1500, the Decay of Nature had almost reached its limit. Taking 4000 B.C. as the date of Creation, Luther calculated that men were already in the sixth and last age of universal history—the Age of the Pope. In principle, this final Age could have lasted until A.D. 2000; but the rot had gone so far that God could surely never permit such utter corruption to continue for another four centuries. 'The world will perish shortly,' he declared: 'The last day is at the door, and I believe the world will not endure a hundred years.'

While few of Luther's contemporaries in Northern Europe were such extreme pessimists, something of his gloom coloured the beliefs and attitudes of all educated men. Everyone assumed that the world was in its last days, though there were disagreements about the precise arithmetic of the Bible chronology. Whereas Luther rounded off the date of Creation to 4000 B.C., other dates also had their advocates—4032, 4004, 3949 and 3946. Johann Kepler, the astronomer, compared the New Testament dating of the Crucifixion with the established cycles of solar eclipses, and detected an error of four years in the chronology of the Christian Era. Evidently, the birth of Christ must be set back to 4 B.C.; and it was on this basis that Archbishop Ussher obtained the date of 4004 B.C. which became the authoritative starting-point of orthodox Anglican chronology.

The new astronomical discoveries of the time only confirmed the widespread pessimism. Tycho Brahe's and Kepler's new stars (of 1572 and 1604) were seen as evidence that corruption had spread from the sublunary world into the very heavens: God had now visited the sins of men on to the whole universe. Galileo's discoveries reinforced these fears. As Sir Thomas Browne—a moderate-minded man, and after his own fashion a scientist—commented:

> When we look for incorruption in the heavens, we find they are but like the Earth; Durable in their main bodies, alterable in their parts; whereof beside Comets and new Stars, perspectives [i.e. telescopes] begin to tell tales. And the spots that wander about the Sun, with Phaeton's favour, would make clear conviction.

As late as the 1630s, we still find Sir Thomas Browne echoing Luther's gloomy prophecies, though he transmutes them into a kind of resigned nostalgia.

> The World grows near its End. . . . The last and general fever may as naturally destroy it before six thousand, as me before forty. [Or

> again:] 'Tis too late to be ambitious. The great Mutations of the World are acted, or Time may be too short for our designes. [Or again:] The World it self seems in the wane, and we have no such comfortable prognosticks of latter times, since a greater part of Time is spun than is to come.

Yet he could find in the very shortness of cosmic history a philosophical consolation that Luther would not have shared:

> He is like to be the best judge of Time who hath lived to see about the sixtieth part thereof [i.e. a hundred years]. In seventy or eighty years a man may have a deep Gust of the World . . . a curt Epitome of the whole course thereof.

In the early seventeenth century, it was a commonplace among English thinkers that the Earth, equally, was corrupted from an original, perfect sphere. Though Milton spoke of the mountains as created by God in all their existing ruggedness, John Donne saw in their peaks and crags yet one more symptom of universal decay:

> But keeps the Earth her round proportion still?
> Does not a Tenarif, or higher Hill
> Rise so high like a Rocke, that one might thinke
> The floating Moone would shipwrack there, and sinke? . . .
> Are these but warts, and pock-holes in the face
> Of th'Earth? Thinke so; but yet confesse, in this
> The world's proportion disfigured is.

It is against this cheerless background that we must consider the new scientific speculations of the seventeenth century. As we shall see, the constricting framework of the older cosmic time-scale did much to distract the 'new philosophers' from questions about cosmic history, and blinded them to the significance of that evidence which could have helped them establish a juster estimate of the world's true antiquity.

Descartes' Mathematical Philosophy

While, generally speaking, the religious tone of sixteenth-century Europe was pessimistic, its dominant intellectual tone was sceptical. Two hundred years earlier, Aristotle had won in science an authority and standing comparable to that of Augustine in theology, and men faced

Nature with the assurance that comes from a convincing body of inherited theory. But now the reimportation of other ancient philosophers into Europe made it clear for the first time just how inconclusive the Greek scientific debate had been. Placed once more alongside their rivals, the theoretical principles of Aristotle's science lost both their magic and their authority. Men found themselves in the position of Socrates: confronted with a chaos of alternative opinions, none of them with any claims to certainty. It was a situation which posed once again, and with the same urgency, the question which Plato had faced—'What, if anything, can be known for certain about Nature?'—and many of the sixteenth-century humanists gave a short answer to this question. Dubious about the very possibility of natural science, they turned instead to literature and the fine arts: the most notable of these sceptics was the French essayist, Michel de Montaigne.

It was this scepticism which the natural philosophers of the early seventeenth century had first to overcome. The two chief prophets of the new scientific era, Francis Bacon and René Descartes, were looking both for an intellectual method which could carry men beyond a barren scepticism, and for a new source of certainty to underpin our ideas about Nature. Bacon put his faith in experience, declaring that experiment and observation alone could provide a reliable basis for scientific ideas. Yet, as Descartes realized, bare facts are never self-explanatory: they require interpreting. Particular observations have to be related back to general principles, and the gravest sceptical doubts arose over these general principles of interpretation. The question was, how could these principles themselves be placed on new and more certain foundations? This was the problem Descartes set himself, and the form of his solution helped to impose on science a fundamentally unhistorical pattern of thought.

Descartes' task of renovation was in the spirit of the times. The Humanists set out to purge the accepted classical texts of corruptions and accretions, hoping to recover authentic Greek and Roman thought. The Protestants similarly tried to purge the Christian religion, and restore the pristine cult of the first centuries A.D. Now Descartes, too, aimed at purging the principles of human knowledge of all but 'clear and distinct ideas'. Only those beliefs which conformed to fundamental axioms whose soundness needed no further demonstration could themselves be regarded as sound: as for the rotten ones, good riddance to them—

> If you have a basket of apples, some of which you know are bad, and are going to spoil and infect the others, the only thing you can do is empty your basket completely and then take and test the apples one by one, putting the good ones back in your basket and throwing away those that are not.

Sixteenth-century humanists like Montaigne had been driven to scepticism about science because they had not been sufficiently *systematic*. Their excursions into natural science had been random and spasmodic, and not one of their theoretical conclusions could stand up to profound intellectual scrutiny. What was needed was a single consistent body of ideas, which would relate all our discoveries together into a coherent system of theory; and, when the intellectual tradition had been purified of obscure and muddled notions, the axioms of the resultant system should be 'self-evident'. As an example of such indubitable knowledge Descartes pointed to the principles of mathematics—which meant for him, as for Plato, the axioms of geometry. Taking these as his starting-point, he went ahead with confidence, building up a theoretical system whose consistency and validity did not depend in any way on revealed truths. As he commented, revelation was concerned with salvation, not with physics, and the mediaeval attempt to expose it to intellectual scrutiny had been largely misguided—'Such a scrutiny could be successful only given some extraordinary aid from heaven: one would have to be super-human!'

The system of physical theory at which Descartes eventually arrived threw little light on the History of Nature. (Its main outlines are explained in chapter 7 of *The Architecture of Matter*: we summarize here only those aspects which concern the temporal development of the natural world.) He had been fifteen years of age when Galileo's *Starry Messenger* reached him at the Jesuit College of La Flèche, and he gladly welcomed its implications. Having rejected the belief that the universe was neatly contained within a finite sphere, he substituted for it an infinite Euclidean space, through which all the different heavenly bodies were scattered more or less randomly. The traditional cosmological image, with its circular boundary, might outlive Descartes, to survive poetically in Milton's *Paradise Lost*, but Descartes could find no reason or authority for confining the world in this way:

> We must beware of thinking too proudly of ourselves. We should be doing this . . . if we imagined any limits to the universe, when none are known to us either by reason or by divine revelation—as if our powers of thought could extend beyond what God has actually made.

Not only were the principles determining the size and layout of the cosmos geometrical, but the laws governing the behaviour of things within it were also mechanical and mathematical. A 'material body' was simply any object having spatial extension, and the principles of mechanics therefore became the laws of 'geometry in motion'. From being an all-embracing organism, animated by intelligences in every part, the cosmos turned into a mathematically-ordered machine: from now on, the womb, entrails, warts and pock-holes of the Earth could no longer be anything but poetic metaphors.

Descartes' faith in mathematics as the fundamental instrument of scientific theory reacted on his attitude towards history. For mathematical propositions were timeless and unchanging, and the laws governing material nature in the present epoch were, accordingly, those which had governed it at all times. From his point of view, the permanence of mathematical principles was one of their chief merits, since they offered him the sure foundation for science which Montaigne had despaired of finding. Descartes thus answered Montaigne in the same way that Plato had answered Socrates: by anchoring down the concepts of science on to timeless, geometrical foundations. But, once having taken this step, he could not allot any serious importance to historical change and development. For him, as for Plato earlier, the temporal flux of historical events lost all fundamental significance, acquiring a theoretical interest only by providing illustrations of the unchanging geometrical laws—the events in question losing, in the process, both their individualities and their dates. 'Histories and fables' were admirable for the same reasons as travel: they were a way of broadening the mind,

> . . . in order to form a sounder judgement of our own [customs], and not think everything contrary to our own ways absurd and irrational, as people usually do when they have never seen anything else.

But serious philosophy did not begin until one saw past the surface flux of events to the geometrical skeleton beneath.

It is true that, in the third and fourth parts of his *Principles of Philosophy*, Descartes did give a speculative account of the way in which the Earth and planets acquired—or *might* have acquired—their present forms. This account describes at some length a mechanical process which would account for their present appearance. For instance—

> Let us therefore imagine that this Earth on which we are was formerly a star composed of the purest matter of the first [self-luminous] element, occupying the centre of one of these fourteen [planetary] vortices . . . and differing from the Sun only in being smaller: but that, as the less subtile parts of its matter joined up bit by bit, they collected together on its surface, and there made up clouds or other thicker and darker bodies, similar to the spots that can be seen on the sun's surface, continually forming and later dissipating; . . . and even that several layers of such bodies were perhaps piled up one above the other, so reducing the strength of the vortex containing [the Earth] that it was

completely destroyed, and that the Earth together with the air and the dark bodies surrounding it descended towards the Sun as far as the place where it is at present.

Descartes' account goes on and on, in elaborate and circumstantial detail, and always in the past tense; so it looks historical enough. But there is a real ambiguity about its status, and, as with Plato's *Timaeus*, we are always compelled to ask: 'How far is this really *meant* as history?' Is he saying, 'This is how it *did* happen', or is he saying only, 'This is how it *could* have happened'? Unfortunately, the complexity of Descartes' motives prevents one from answering this question with any certainty.

If we take Descartes at his written word, the historical aspect of his geology and cosmology must be dismissed as illusory. All those apparently direct assertions about the astronomical and geological past are, in fact, only samples from an indefinite range of intellectual fictions:

> In order to explain natural objects the better, I shall pursue my inquiry into their causes further back than I believe the causes ever in fact existed. There is no doubt that the world was first created in its full perfection; there were in it a Sun, an Earth, a Moon, and the stars; and on the Earth there were not only the seeds of plants, but also the plants themselves; and Adam and Eve were not born as babies, but made as full-grown human beings. This is the teaching of the Christian faith; and natural reason convinces us that it was so; for, considering the infinite power of God, we cannot think he ever made anything that was not perfect.
>
> Nevertheless, in order to understand the status of a plant or man, it is far better to consider how they may now gradually develop from seeds, rather than the way they were created by God at the beginning of the world; and in just the same way we may conceive certain elements, very simple and very easily understood, and from these seeds (so to say) *we may prove that there could have arisen* stars, and an Earth, and in fact everything we observe in this physical universe; and although we know perfectly well *they never did arise in this way*, yet by this method we shall give a far better account of their nature than if we merely describe what they now are.

Yet are these scruples really sincere? Or is he merely being politic, and inserting the diplomatic disclaimers needed in order to cover him against charges of religious infidelity?

These questions are not easily answered. Descartes might well have entered some such reservations for straightforward philosophical reasons. The ambiguities inherent in the idea of Creation were familiar enough to his contemporaries: Sir Thomas Browne, for instance, remarked in his *Religio Medici*—

> Some believe there went not a minute to the World's Creation.... Those six days rather seem to manifest the method and Idea of the Great Work of the Intellect of God, than the [historical] manner how He proceeded in its operation.

At times, Descartes' phrasing suggests that this was indeed his attitude—

> We may conceive certain elements from which there could have arisen everything we observe; and by this method we shall give a far better account of their nature than if we merely describe what they now are.

Yet his reservations represent only a minute fraction of the whole *Principles of Philosophy*. For the rest, he writes so freely and circumstantially about 'the origins of the material world' that one finds it difficult to believe that he is actually suspending judgement.

Here extraneous suspicions also become relevant. Descartes had grave reasons for presenting his revolutionary ideas as innocently and undogmatically as possible. His was an age of ideological pressures. While he was actually writing a first account of his physical system, the news reached him of Galileo's trial. The condemnation by the Inquisition of the man whom he so much admired left a deep impression on him. As he explained three years later:

> I was beginning to revise my *De Mundo* [this account] so as to put it into the hands of a printer, when I learned that certain persons to whom I defer, and who have hardly less authority over my actions than my own reason has over my thoughts, had disapproved of a physical theory published a little while before by somebody else. I will not say I held this theory, but I had certainly noticed nothing in it, before their condemnation, that I could imagine prejudicial to either religion or society; nothing, therefore, would have stopped me putting it in writing if reason had convinced me of it.
>
> This made it clear that there might be some mistake in my own theories, in spite of the great pains I have always been at not to admit to belief in any new ones of which I had not very certain demonstration, and not to write anything that could turn out to the disadvantage of anybody; and it was enough to alter my previous decision to publish.

If Galileo was a Pasternak, Descartes was a Yevtushenko. He had always known the virtues of discretion. In a private notebook started at the age of twenty-two, he confessed to himself, 'Now that I am to mount the stage of the world, where hitherto I have been a spectator, I come forward *wearing a mask.*' Rather than defy ecclesiastical authority, he disengaged from the threatened conflict, rewrote the account of his theory and withheld it from publication for ten years. Consequently, when his *Principles* were finally issued, many of his contemporaries treated all those careful qualifications, about Adam and Eve being 'made as full-grown human beings', and the rest, as so much eyewash.

Still, there is one further possibility. Perhaps Descartes wrote as he did from motives which—while mixed—were at the same time sincere. For once in a while, logical and religious scruples pointed quite genuinely in the same direction. The man whose fundamental principle was to call everything in question, and who vowed to retain only those beliefs he could establish unanswerably, could hardly have rated his own account of the origins of the world-machine as 'the actual truth'. He must have known, better than anyone, that his theory was little better than guesswork. So, if his mechanical history of the cosmos was intended as a call-to-arms, the trumpet-blast was decidedly muted; and it took almost two centuries for cosmology and geology to divorce themselves finally from *Genesis.*

The Theological Consequences of Cartesianism

Behind Descartes' ambiguous attitude to history lies a deeper ambiguity in his theological attitudes. On the one hand, he had felt for himself the full force of Montaigne's sceptical arguments; yet at the same time his own education, from ten to eighteen, had been a religious one. This duality coloured all his theories. While Montaigne had inoculated him against naive assumptions about Providence—e.g. that the Sun, Moon and stars 'were established and continued so many ages for [Man's] commodity and service'—there is not much doubt that throughout his life Descartes, like Galileo, considered himself a sincere and devout Christian.

From one point of view (it is true) the whole Cartesian system of natural philosophy appeared 'a-religious', for throughout it God was (so to speak) held at arm's length. The explanations Descartes gave drew their immediate authority from the reliability of the human reason—

> We have thus discovered certain principles as regards material objects, derived not from the prejudices of our senses but from the light of reason, so that their truth is indubitable; we must now consider whether they suffice to explain all natural phenomena.

The key historical events which had hitherto represented Divine interventions in history—the Creation, the Flood and the final Conflagration—were swept out of physical cosmology: in their place, Descartes offered men the self-sufficient development of a world-machine, operating continuously according to self-evident principles. Those of his readers who had hitherto seen the mysterious purposes of God in the smallest operations of nature were thrilled—or horrified—at his suggestion that, after all, the material world might be both self-sufficient and completely open to the human understanding.

Yet the system could be viewed from the other side: the Almighty was the only guarantee men had that their reason was not deceiving them. So in the last resort, the very certainty at which Descartes aimed still depended on a belief and trust in the Creator.

> Assuredly, if the only principles we use [in our physical theories] are such as we see to be self-evident; if we infer nothing from them except by mathematical deduction; and if these inferences agree accurately with all natural phenomena; then we should, I think, be wronging God if we were to suspect this discovery of the causes of things to be delusive. God would, so to say, have made us so imperfectly that by using reason rightly we nevertheless went wrong.

Descartes' Protestant opponents saw this reference to God as mere lip-service—an afterthought introduced at the last moment, to save an essentially atheistical system from its just condemnation. However there is little doubt that at this one fundamental point Descartes was quite deliberately giving his vision of Nature a religious basis. Though the Hand of God was not to be seen in particular details of the natural world, none the less it was He who had endowed the human mind with the rationality needed for interpreting Nature.

The sceptics and rebels who frequented the salons of seventeenth-century Paris considered that Descartes had been needlessly cautious, and welcomed the Cartesian system as much as the Church deplored it. They were men of very different temperament from Descartes himself and, instead of evading the inconsistencies between *Genesis* and the mechanical world-system, they applauded them. Descartes' suppositions became their assertions. Disregarding the genuflections which qualified his cosmology, they began to treat his theory of vortices as a plausible, likely—even as an established—theory of the Origin of the World.

This radical Cartesianism culminated in the most brilliant work of scientific popularization ever witten, Bernard de Fontenelle's *Conversations on the Plurality of Worlds*. Fontenelle presents his story as a sequence of five evenings, which he spends walking with an imaginary Marquise in her park, while explaining to her the latest ideas about the universe. With a mixture of wit, urbanity and logic, he persuades her first, that the Earth rotates once a day, then that it travels round the Sun once a year, and next that—for all we can tell—both the Moon and the other planets have inhabitants of their own: scatter-brained and 'mercurial' on the hot planets nearer the Sun, slower-witted and more 'saturnine' on the colder and more distant ones. He goes on to discuss the difficulties in the way of interplanetary travel. Every planet or satellite, for example, has its own distinct atmosphere—

> 'Water is the air of fishes; they never pass into the air of birds, nor the birds into the air of fishes; it is not the distance which hinders them, it is, that each hath for a prison the air it respires. . . . An inhabitant of the *Moon*, who should arrive on the confines of our World, would be drowned the moment he entered into our air, and we should see him fall dead on the *Earth*.'
>
> 'O! what a desire have I,' cried the *Marquise*, 'that there might happen some great shipwreck which would scatter here a great number of these people, then we might consider at ease their extraordinary figures.'
>
> 'But,' replied I, 'if they should be able to swim on the exterior surface of our air, and from thence through a curiosity to see us, should fish for us as we do for fishes, would this please you?'
>
> 'Why not?' answered she, laughing. 'As for me I would readily throw myself into their nets, only to have the pleasure of seeing those who fish for me.'

Yet all this fancy is only a preparation for the most radical implications of the Cartesian doctrine—that space is *infinite*:

> 'If I am not deceived,' said the *Marquise*, 'I already see to what you are leading. You are going to say the Fixed Stars are so many Suns, our Sun is the centre of a Vortex, which turns round him, so why therefore shall not each Fixed Star be also the centre of the Vortex, which shall have a motion round it? Our Sun hath Planets that he enlightens, so why therefore should not each Fixed Star have Planets that he enlightens?'
>
> 'I have not anything to answer,' said I to her, 'than that which Phaedrus said to Enone, *It is thee who hath named it*.'
>
> 'But,' replied she, 'I see the Universe so great, that I am lost in it. I no longer know where I am: I no longer know anything. What, shall all be divided into vortexes thrown confusedly one amongst another? Shall each Fixed Star be the centre of a vortex, as great

perhaps as that in which we are situated? Shall all this immense space, which comprehends our Sun and our Planets, be only a little parcel of the universe? Shall there be as many such spaces as there are Fixed Stars? This confounds me, troubles me, frights me.'

'And as for me,' answered I, 'I am very easy about it.'

The Marquise eventually becomes reconciled to this indefinite plurality of planetary systems, each with its own inhabitants, only to receive another, greater shock:

> 'What!' cried she, 'are Suns extinguished?'
>
> 'Yes, without doubt,' answered I. 'The Ancients saw in the Heavens certain Stars which have disappeared, and never been seen since' . . .
>
> 'You make me tremble,' said the *Marquise*. 'Now that I know the consequences that may happen from the Sun's paleness, I believe, that instead of going in the morning to see in my glass, whether I am pale or not, I shall go and look in the Heavens to see if the Sun is pale.'
>
> 'Ah! Madam,' answered I, 'be assured, it will require a long time to ruin a World.'
>
> 'But it may in the end,' replied she. '*It only requires time.*'

From the newly-accepted doctrine that the world was infinitely extended in space, we can here see emerging the first hints of the complementary idea—that its span of existence would be, and perhaps already had been, equally unbounded in time. For the moment, there were very few who could appreciate the force of this parallel, let alone pursue it. Nor was there empirical evidence available which could impose an extension of the time-scale in the way Galileo's observations had done for space. Still, the new 'mechanical philosophy' did achieve one immediate effect. The phantom which had haunted the minds of Christians for centuries, of the imminent destruction of the whole cosmos, at last began to fade. Presumably the mathematical principles underlying Nature could not be disrupted by human sin, and would continue to operate unchanged for the foreseeable future. So the intellectuals of Europe, who a hundred years earlier had been debating whether Noah's Ark could possibly have accommodated two of every known species, were given a criterion for separating scientific questions from religious ones—for keeping the problems of geology and astronomy apart from those of Biblical interpretation.

In France itself, there were many who took up this opportunity gladly. During the next century, scientists in France became more and more sceptical about divine interventions in the affairs of the natural world. The Creator had had enough foresight to make Nature self-sufficient. Most people still assumed that God intervened in human affairs, but the influential group of Deists denied even that. They restricted God's role in the world to the initial Creation: in the beginning, He had set the cosmic machine working according to fixed laws, and from then on it had gone its own way.

The Blueprints of Creation

The majority of seventeenth-century scientists, however, were neither Cartesians nor Deists, and they saw the intellectual situation in a very different light. Descartes' proposed divorce between science and theology was far too extreme for them: the fundamental principles of his physics removed the Hand of God entirely from the actual course of natural events, and this appeared a scandal. Protestant scientists, in England especially, dismissed the Cartesians scornfully as a pack of 'mechanick theists'. Not only were they determined to preserve a central place for God within the scientific view of the natural world: they saw in every aspect of Nature positive evidence of His continuing participation and all-embracing wisdom. They agreed with Descartes in seeing the operations of Nature as resting on mechanical or mathematical principles; but in their eyes these principles were, not the self-sufficient products of the human reason, but the actual blueprints to which the Creator worked. For Descartes, the Laws of Nature formed a self-justifying logical system: for John Ray and Isaac Newton, they expressed the choice and intention of God.

Since in studying the world of Nature the devout scientist was deciphering the Plan of Creation, he thereby came closer to understanding the mind of God. So he had no reason for keeping God at arm's length, as Descartes had done. Rather the opposite: for him, science and theology were two aspects of the same thing, and to divorce them would be to emasculate science itself. This view not only gave men a supreme motive for studying Nature: it also inclined them to fundamentalism. To question the word of God, as expressed in the Bible, would be to contradict the very faith which lay at the heart of their own scientific studies.

It is against this background that we must judge the arguments about the History of Nature which were to occupy the next two hundred years. Only with this in mind can we appreciate why the tone of these conflicts was so bitter, and why the task of interpreting the empirical evidence, as it became available, involved men in such agonizing intellectual reappraisals. For the conflicts were not between science and religion: they were *within science*, as men then conceived it—not the struggles of progressive-minded scientists to overcome the obscurantism of an external authority, but internal schisms within the new and highly successful philosophy.

So, in the years following 1650, men were still constrained within the framework of their old beliefs, and these included the traditional time-scale—however elastically it might be interpreted. The scientists might be observant, perceptive and clear-headed, but at some point their interpretations inevitably came up sharp against Biblical history, and the resulting strains can be seen in every field of science. The men concerned were too scrupulous to ignore them. When, in old age, John Ray scrutinized the fossil traces of extinct ferns, he found his scientific honesty at odds with the religious preconceptions which had guided his thoughts hitherto: as he confessed in a letter to a friend,

> ... there follows such a train of consequences, as seem to shock the Scripture-History of ye novity of the World; at least they overthrow the opinion generally received & not without good reason, among Divines and Philosophers, that since ye first Creation there have been no species of Animals or Vegetables lost, no new ones produced.

With the same honesty he drew attention, in another private letter, to signs of a geological antiquity which no one had yet the courage to affirm publicly—

> I gather that all the other vast Stones that lie in our mountainous Valleys [in Wales] have by such accidents as this fallen down: Unless perhaps we may do better to refer the greatest part of them to the universal Deluge. For considering there are some thousands of them in these two valleys . . . whereof there are but two or three that have fallen in the Memory of any Man now living; in the ordinary Course of Nature we shall be compelled to allow the rest many thousands of years more than the Age of the World.

Geology

It was here that the shoe first began to pinch. Consider the two leading geologists of the seventeenth century—the Englishman, Robert Hooke, and the Dane, Nils Steensen, usually known by the latinized name of Steno. Like Leonardo da Vinci before them, both men recognized the crucial significance of organic fossils. They were not content, as so many of their contemporaries were, to collect fossils merely as curious objects—to be admired, rather than understood—nor to let ignorance take the easy way out, of treating them as 'sports of Nature' generated spontaneously within pre-existing rocks. Instead, they acknowledged that in all probability fossils were formed at the same time as the rocks in which they were found, and were the petrified traces of organisms comparable to those of the present day.

Starting from this point, Hooke and Steno braced themselves to face the intellectual consequences, and carried their arguments remarkably far. Hooke realized immediately that the presence of marine fossils in the rocks of inland mountains implied unsuspected alterations in the Earth's

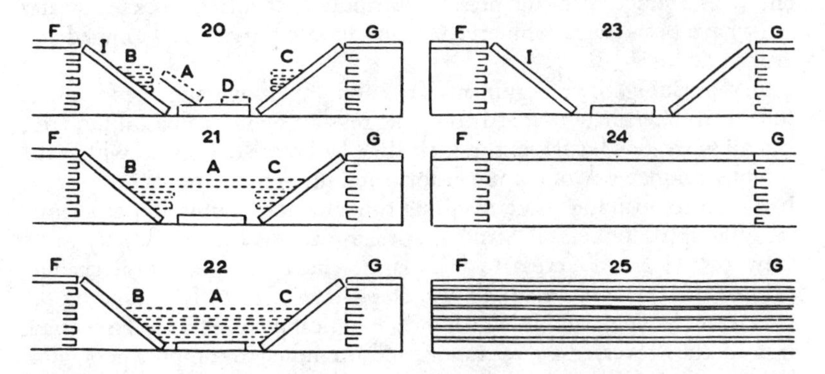

Steno's explanation of the successive stages by which stony beds are eaten out 'by the force of Fire and Water', to form valleys and mountains 'by the ruine of the Superior Beds': from the first English edition of his *Prodromos* (1671). NB: The sequence begins with no. 25 and ends with no. 20

crust. The fossil evidence could be explained readily enough, if one only supposed that:

> A great part of the Surface of the Earth hath been since the Creation transformed and made of another Nature; namely, many Parts which have been Sea are now Land; and diverse other Parts are now Sea which were once a firm Land; Mountains have been turned into Plains and Plains into Mountains and the like.

Erosion by wind and rain, the action of rivers, floods and the sea, gravity dragging the rocks and soil downwards, earthquakes and eruptions thrusting them up again—there seemed no limit to the geological transformations which could be brought about by these natural agencies.

Steno, for his part, carried the analysis still further. Travelling around Tuscany, he observed the conformation of the hills and valleys, and the directions of the rock-strata at each level; and, by postulating a sequence of distinct geological phases, he was able to explain (in 1669) how the area had acquired its present appearance. In many respects, his argument was entirely modern. Any stratum containing marine fossils must originally have formed part of the bed of a former sea, and so have been horizontal. Having been lifted up violently to form an elevated plateau, the stratum was then undermined by subterranean forces and subsequently broke apart, to create a valley with sloping strata on either side. These strata were subsequently worn down, to become in due course part of a fresh sea bed, in which further marine fossils were deposited, and so on. . . . To judge from the present distribution of surface rocks, Tuscany must have been twice submerged, twice elevated, twice undermined and broken down.

By pursuing these arguments far enough, Hooke and Steno might (one feels) have inaugurated a complete revolution in geological thought. But all scientific speculations in their time had to be conceived within the accepted framework of cosmic history, and this left geologists little scope. It is hard to abandon more than one fundamental assumption at a time: so, although Hooke and Steno saw their way past an absolutely static view of the Earth's crust to a dynamic theory of geological change, neither of them could take the next step—that of forcibly expanding the overall chronology within which the geological picture had to be framed. Instead, they reversed the operation, and attempted to compress geological history within the accepted time-scale; and the intellectual strains created by this compromise explain why their ideas had such little immediate influence.

They met with two embarrassments in particular, and these reinforced one another. On the one hand, to produce the required effects in the time available, the cataclysms responsible for shaping the Earth's surface must have been vastly more drastic and frequent than those we experience nowadays—the earthquakes more formidable, the floods more engulfing. On the other hand, the available histories mentioned only one such catastrophe since the Creation, *viz.* the Biblical deluge. The first of these embarrassments was the ruination of Hooke and Steno's intellectual method. The force of their arguments had depended initially on extending to former ages their knowledge of the forces shaping the Earth's surface in modern times. (To that extent they were 'uniformitarians'—seeing geological change as the product of the same natural agencies acting at all times in similar ways.) But, when faced with the need to compress the Earth's history, Hooke and Steno went back on their tracks, and concluded that in earlier times the Earth's crust was far more malleable and subjected to more violent forces.

Why, then, were these geological catastrophes undocumented? Again, the truth was greater than they could foresee. Taking it for granted that the human species was almost as old as the Earth, Robert Hooke combed the literature of classical antiquity for allusions to those vast geological events which his reasoning convinced him had taken place. He found some encouragement in the myth of Atlantis, but ended by placing most faith in Ovid's *Metamorphoses*: this he read as a mythological account of the creation and formation of the world. Steno also read geological lessons into ancient history and legend. The fossilized elephant-bones he found in Tuscany were presumably relics of Hannibal's army, and the difficulty scholars met when reconstructing the journeys of Ulysses and Aeneas proved how much Mediterranean geography had changed since their day.

Robert Hooke did face one other important fact. If fossils were truly petrified organic remains, then many species in earlier ages consisted of 'creatures, that are now quite lost, and no more of them surviving on any part of the earth'. This radical suggestion, that organic species differed from epoch to epoch, chimed in nicely with his views on geological mutability. Fossils recovered at Portland, on the South coast of England, resembled those of modern tropical species, while elsewhere the fossil creatures seemed often to have been 'of a much greater and gigantick standard; suppose ten times as big as at present'. Evidently, in Ovid's Golden Age, climate and geography had a perfection since lost, and supported larger and stronger species than the world knows today. More recently, Hooke inferred, the original supple skin of the Earth had dried out and fallen apart, leaving the battered ruin characteristic of the present 'Iron Age', while living species had 'dwindled and degenerated into a dwarfish progeny'.

Operating within the restrictive framework of Biblical chronology, geological observation and inference seemed only to confirm the prejudices of seventeenth-century pessimism. Hooke convinced himself of terrestrial and zoological change, by extrapolating familiar processes back into the remote past. But he could not carry this 'uniformitarian' programme through to the crucial point at which the traditional time-scale had to be abandoned. Rather, he attributed the most drastic geological changes (such as Noah's Flood) to 'a preternatural *digitus Dei*'. Steno, for his part, having made his one major contribution, set geology aside, was converted to Catholicism, and in a few years was appointed Bishop. The mechanisms of geological change no longer interested him: instead, he found in his discoveries evidence of 'the agreement of Nature with Scripture', and 'congratulated himself [as Leibniz reported] with having come to the aid of piety in supporting the faith of the Holy Scriptures and the tradition of universal deluge on natural proofs'.

This phase in geological speculation culminated in Thomas Burnet's *Sacred Theory of the Earth*, whose four laborious volumes all appeared within a few years of Newton's *Principia*. To educated men in the 1680s,

Burnet's work was the most outstanding—or outrageous—product of modern thought. It was the first thorough attempt to retell the Biblical story of the world in terms of the new discoveries of seventeenth-century physics. Burnet was convinced that everything in Holy Scripture (the *words* of God) must be consistent and reconcilable with everything in Nature (the *works* of God), and the grandiosity of his intentions was matched by a literary style which was vivid, and even lurid. Whereas Hooke and Steno had appealed to supernatural interventions only to explain what natural science could not, Burnet did something far more shocking. With all the literary power at his command, he presented an account of the world's history in which all the traditional Bible stories, and even the Hand of God itself, were re-interpreted in naturalistic terms. Before the Flood (for instance) the world had a charm and tranquillity which are reflected in Burnet's prose:

> In this smooth Earth were the first Scenes of the World, and the first Generations of Mankind; it had the Beauty of Youth and blooming Nature, fresh and fruitful, and not a Wrinkle, Scar or Fracture in all its Body; no Rocks nor Mountains, no hollow Caves, nor gaping Channels, but even and uniform all over. And the Smoothness of the Earth made the Face of the Heavens so too; the Air was calm and serene; none of those tumultuary Motions and Conflicts of Vapours, which the Mountains and the Winds cause in ours; 'Twas suited to a golden Age, and to the first innocency of Nature.

In the time of Noah, however, Providence 'thought fit to put a Period' to this first delicious condition. The original smooth Earth was ravaged by the Flood, which scoured its surface into a novel ruggedness; and at this point a rough abruptness enters Burnet's pen.

> Thus the Flood came to its height; and 'tis not easy to represent to our selves this strange Scene of Things, when the Deluge was in its Fury of Extremity; when the Earth was broken and swallowed up in the Abyss, whose raging Waters rise higher than the Mountains, and fill'd the Air with broken Waves, with an universal Mist, and with thick Darkness, so as Nature seem'd to be in a second Chaos; and upon this Chaos rid the distress'd Ark, that bore the small Remains of Mankind . . . a Ship whose Cargo was no less than a whole World; that carry'd the Fortune and Hopes of all Posterity, and if this had perish'd, the Earth for any thing we know had been nothing but a Desart, a great Ruin, a dead heap of Rubbish, from the Deluge to the Conflagration. But Death and Hell, the Grave and Destruction have their Bounds.

The reader sighs with relief. Weary and seasick, the occupants of the Ark climb out on to dry land, to find themselves among 'the Ruins of a broken World'—but at any rate they had come through. . . . And so Burnet's tale goes on, embracing in its scope the final Conflagration, and the New Heaven and New Earth prophesied in the *Book of Revelation*.

Burnet's *Sacred Theory* reads more like science fiction than serious geology, and the French naturalist Buffon described it somewhat ambiguously as 'a fine historical romance'. Yet the book was highly influential and attracted as much attention as Newton's *Principia* itself. Orthodox theologians found it blasphemous: the greater Burnet's efforts to harmonize Scripture and geology, the more he drew attention to their disharmony and exposed the religious tradition to fresh embarrassments. Modernist theologians, on the other hand, saw in it the way to remove these same embarrassments, and to establish the literal authority of the Bible in the face of the Cartesian challenge. Burnet's self-confident manifesto freed Protestant thinkers from their earlier inhibitions far more than did the tentative speculations of Hooke and Steno, and set them speculating about the natural development of our globe. Here again, scientific research became a pious duty, enabling men to recognize the divine handiwork in the evidences of Nature. The cosmic drama had left its mark on the Earth's crust, and intelligent geological observation would reconstruct its acts and scenes. For much of the eighteenth century, Burnet's book remained a best-seller, and its ideas were still current well after 1800: Coleridge admired the 'Tartarean fury and turbulence' of Burnet's style so warmly that he even contemplated turning parts of the *Sacred Theory* into a blank-verse epic!

Cosmology

The strains involved in adapting science to the Scriptural time-scale can be seen in planetary theory also. In his laws of motion and gravitation, Isaac Newton seemed to have demonstrated with a new profundity the Creator's plan for Nature. Yet he took care to distinguish sharply between the established Order of Nature, 'created by God at first, and by Him conserved to this Day in the same State and Condition', and the original act of Creation by which that Order was instituted. The tidy arrangement and regular motions of the planetary system did not 'arise out of a Chaos by the mere Laws of Nature; though being once form'd it may continue by those Laws for many Ages'. Descartes might be willing to speculate about a mechanical origin for the planetary system: Newton was not. Instead, he followed Galileo in supposing that the planets, having fallen from a great distance in straight lines, had been switched by God into circular orbits around the Sun, though he pointed out that

> The divine power is here required in a double respect, namely to turn the descending motions of the falling planets into a side motion and, at the same time, to double the attractive power of the sun.

However wild and unverified Descartes' speculations had been, he had opened doors through which one might glimpse a more dynamic view of Nature. Newton turned his back on that vision, preferring the static picture advocated by his friend, Robert Boyle, in the words:

> The philosophy I plead for, reaches but to things purely corporeal; and distinguishing between the first origin of things and the subsequent course of nature, teaches, that God, indeed, gave motion to matter; but that, in the beginning, he so guided the various motions of the parts of it, as to contrive them into the world he design'd they should compose; and established those rules of motion, and that order among things corporeal, which we call the laws of nature. Thus, the universe being once form'd by God, and the laws of motion settled, and all upheld by his perpetual concourse, the general providence; the same philosophy teaches, that the phenomena of the world, are physically produced by the mechanical properties of the parts of matter; and, that they operate upon one another according to mechanical laws.

The Creation itself was not an event within cosmic history. Only the subsequent operations of Nature, governed by the laws established at the Creation, could be interpreted in mechanical terms.

Yet, in the long run, the success of Newton's physical theories helped to promote the very views he had himself rejected. He had demonstrated in his *Principia* that Descartes' theory of vortices could not be made to explain the actual motions of the planets without the most implausible and elaborate assumptions. But it was one thing to refute the details of Cartesian physics, and quite another to discredit its fundamental programme. That Newton never succeeded in doing. Eighteenth-century cosmologists cheerfully jettisoned Descartes' 'vortices' in favour of Newton's 'forces', and then carried on as before—in a strictly Cartesian spirit. By exploiting Newton's successes, they were better able to realize Descartes' fundamental ambitions.

Newton's successor at Cambridge, William Whiston, tried for a while to establish detailed correspondences between physical cosmology and the Old Testament. He treated the Biblical Flood as the gravitational by-product of a passing comet, which drew up all the waters from the surface of the Earth and submerged the dry land. (By a combination of astronomical and Scriptural considerations, he dated this event to 27th November 2349 B.C.) But the days of such compromises were numbered. As Buffon remarked in 1749, Whiston's attempt to weave a consistent fabric out of Scripture and physics 'so strangely jumbled divinity with human science, that he had given birth to the most extraordinary system that perhaps ever did or ever will appear'.

Buffon himself insisted that the only way of escaping these paradoxes was to emancipate cosmology entirely from the *Book of Genesis*:

> Whenever men are so presumptuous as to attempt a physical explanation of theological truths; whenever they allow themselves to interpret the sacred texts by views purely human; whenever they reason concerning the will of the Deity and the execution of his decrees; they must necessarily involve themselves in obscurity and tumble into a chaos of confusion.

This emancipation could be achieved, he argued, by carrying Newton's theories one stage farther. Newton had remarked that the planets all travel round the Sun in the same direction and in roughly the same plane: this fact had been for him one more evidence of the divine handiwork. Yet, as Buffon pointed out, this uniformity could equally well be explained mechanically, using Newton's own ideas. Taking his cue from Whiston, he supposed that a comet had passed close by the Sun, drawing off from its surface gravitationally about 1/650th part of its bulk, in hot and liquid form. Once detached, this matter would have condensed into a number of much smaller bodies, and these would have cooled more rapidly than the Sun, while continuing to rotate around it in the same plane at greater or lesser speeds. Meanwhile, the retreating comet might have struck off still smaller portions from the newly-forming planets, and these too would continue moving in the same plane as the planets, as their satellites.

This theory, which ignored theology entirely, roused a storm before which Buffon was obliged to bow. Threatened with censorship, he was compelled to retract the theory, and put his private thoughts into cold storage. But censorship cannot kill heterodox ideas. In 1750, one year after Buffon had published his own theory, Thomas Wright of Durham carried the argument still farther: recognizing that the Milky Way was a cross-section of a flat galaxy, he speculated that the whole stellar universe might have been formed by the action of gravity. This was only a beginning. As we shall see, five years later Immanuel Kant was introducing a novel, evolutionary dimension into cosmology, and the devout Newtonian attitude to the divine Creation was being left far behind.

Meanwhile, the general pattern characteristic of the 'new philosophy' was leaving its stamp on other fields of study. We have seen how Christian scholars fitted Aristotle's Ladder of Nature into the mediaeval world-allegory, so transforming a zoological sequence of organisms into a religious hierarchy of created beings. In the years following 1650, this conception experienced a fresh metamorphosis, which fitted it to the intellectual demands of the new age. Throughout this period the dominant current of thought was flowing steadily in a constant direction. God the Supreme Commander, supervising the affairs of the material world, was everywhere being superseded by God the Supreme Designer, contemplating the operation of His world-machine, and intervening only in man's spiritual life by occasional acts of Grace.

This new framework left little scope for the mediaeval interpretation of the Ladder of Nature, as a 'chain of command' linking together all created things. Yet, in its new form, the idea remained as influential as ever, and it played a crucial part in setting the stage for nineteenth-century debates about evolution. Indeed, some ideas which in retrospect look specifically Darwinian appeared on the scene a century earlier, in a non-evolutionary context. The most striking example is the idea of 'the missing link' between rational and non-rational creatures—those hypothetical primates, intermediate between the most intelligent known great apes and the most stupid human beings. We think of these creatures now—if at all—as the ancestors of man, intervening in the family-tree between the existing human species and the extinct anthropoid apes. But this was not their original standing. In fact, the very name 'missing link' places them securely in a pre-Darwinian context, for the chain in which they belonged was the one known to the eighteenth century as the Great Chain of Being.

The doctrine of the Great Chain of Being is described by John Locke in his *Essay concerning Human Understanding*:

> In all the visible corporeal world we see no chasms or gaps. All quite down from us the descent is by easy steps, and a continued series that in each remove differ very little one from the other. There are fishes that have wings and are not strangers to the airy regions; and there are some birds that are inhabitants of the water, whose blood is as cold as fishes. . . . There are animals so near of kin both to birds and beasts that they are in the middle between both. Amphibious animals link the terrestrial and aquatic together; . . . not to mention what is confidently reported of mermaids or sea-men. There are some brutes

> that seem to have as much reason and knowledge as some that are called men; and the animal and vegetable kingdoms are so nearly joined, that if you will take the lowest of one and the highest of the other, there will scarcely be perceived any great difference between them; and so on until we come to the lowest and the most unorganical parts of matter, we shall find everywhere that the several species are linked together, and differ but in almost insensible degrees.
>
> And when we consider the infinite power and wisdom of the Maker, we have reason to think, that it is suitable to the magnificent harmony of the universe, and the great design and infinite goodness of the architect, that the species of creatures should also, by gentle degrees, ascend upwards from us towards his infinite perfection, as we see they gradually descend from us downwards.

This conception dominated eighteenth-century thought to an extent which it is hard to appreciate today. From Leibniz and Locke, through Addison, Bolingbroke and Pope, Buffon and Diderot, to Kant, Herder and Schiller; one after another, one finds the most influential eighteenth-century authors accepting this notion unquestioningly. In an unlikely alliance, Voltaire and Samuel Johnson were almost the only sceptics.

The Great Chain of Being extended from the highest beings to the lowest. With Man as starting-point, one could proceed up or down:

> The Ape or the Monkey that bears the greatest Similitude to Man, is the next Order of Animals below him. Nor is the Disagreement between the basest of Individuals of our species and the Ape or Monkey so great, but that, were the latter endow'd with the Faculty of Speech, they might perhaps as justly claim the Rank and Dignity of the human Race, as the savage *Hottentot*, or a stupid native of Nova Zembla. . . .
>
> The most perfect of this Order of Beings, the *Orang-Outang*, as he is called by the natives of *Angola*, that is the Wild Man, or the Man of the Woods, has the Honour of Bearing the greatest Resemblance to Human Nature. Tho' all that Species has some Agreement with us in our Features, many instances being found of Men of Monkey Faces; yet this has the greatest Likeness, not only in his Countenance, but in the Structure of his Body, his Ability to walk upright, as well as on all fours, his Organs of Speech, his ready Apprehension, and his gentle and tender Passions, which are not found in any of the Ape Kind, and in various other respects.

Jean-Jacques Rousseau, in fact, lumped humans, orang-outangs and chimpanzees together in a single species, treating speech as an acquired human capacity rather than as a specific characteristic. The Scottish author and landowner, Lord Monboddo, went even farther. He actually contemplated educating orang-outangs, and hoped that they might eventually take over the more menial forms of housework. In this way, he drew down on his head Dr. Johnson's characteristically heavy-handed witticisms—

> *Edinburgh, 16 August 1773* We talked of the *Ouran-Outang*, and of Lord Monboddo's thinking he might be taught to speak. Dr. Johnson treated this with ridicule. Mr. Crosbie said, that Lord Monboddo believed the existence of everything possible; in short, that all which is in *posse* might be found in *esse*—Johnson: 'But, Sir, it is as possible that this Ouran-Outang does not speak, as that he speaks. However, I shall not dispute the point. I should have thought it not possible to find a Monboddo; yet *he* exists.'

At the lower extremity of the Chain, the microscopical discoveries of Antony van Leeuwenhoek had filled the gap separating the smallest visible organisms from the inanimate world. Addison referred to these animalculae and protozoa in the *Spectator*:

> Every part of Matter is peopled; every green Leaf swarms with Inhabitants. There is scarce a single Humour in the Body of a Man, or of any other Animal, in which our Glasses do not discover Myriads of living Creatures. The Surface of Animals is also covered with other Animals, which are in the same manner the Basis of other animals that live upon it; nay, we find in the most solid Bodies, as in Marble itself, innumerable Cells and Cavities that are crouded with such imperceptible inhabitants, as are too little for the naked eye to discover.

Going back to Man and proceeding upwards, further 'missing links' were presumed, superior to Man but inferior to God. Here is Addison again:

> If the notion of a gradual rise in Beings from the meanest to the most High be not a vain imagination, it is not improbable that an Angel looks down upon a Man, as a Man doth upon a Creature which approaches the nearest to the rational Nature.

So far as the visible inhabitants of the Earth went, Man was the highest form known, but astronomy promised to make good the deficiencies of terrestrial zoology. Fontenelle's arguments for a plurality of inhabited worlds had carried general conviction: accordingly, one might believe either that the superior beings were incorporeal and so invisible, inhabiting interplanetary space, or alternatively that the multitude of stars included some whose planets were inhabited by super-men. Kant speculated that creatures living on the planets closer to the Sun would be more corporeal and so less spiritual than terrestrial beings, and the opposite would be true for those whose home was farther away. Since the more distant planets received less heat and light, intelligent life would be possible only for organisms composed of a 'lighter and finer' matter:

> The excellence of thinking natures, their quickness of apprehension, the clarity and vividness of their concepts, which come to them from the impressions of the external world, their capacity to combine these concepts, and finally, their practical efficiency, in short the entire extent of their perfection, becomes higher and more complete in proportion to the remoteness of their dwelling-place from the Sun.
>
> Human nature occupies as it were the middle rung of the Scale of Being, . . . equally removed from the two extremes. If the contemplation of the most sublime classes of rational creatures, which inhabit Jupiter and Saturn, arouses his envy and humiliates him with a sense of his own inferiority he may again find contentment and satisfaction by turning his gaze upon those lower grades which, in the planets Venus and Mercury, are far below the perfection of human nature.

To students of English literature, the best-known statement of this eighteenth-century doctrine is contained in Alexander Pope's *Essay on Man*. Pope's uncritical use of the stable Order of Nature to justify political conservatism is at times infuriating—

> Order is Heav'n's first law; and this confess'd,
> Some are, and must be, greater than the rest,
> More rich, more wise . . .

Nevertheless, in a famous passage which records the ambiguity of Man's position at the mid-point of the Great Chain, he does succeed in capturing something of the compelling grandeur of the eighteenth-century vision:

> Plac'd in this isthmus of a middle state,
> A being darkly wise and rudely great,
> With too much knowledge for the sceptic side,
> With too much weakness for the stoic pride,
> He hangs between; in doubt to act or rest;
> In doubt to deem himself a god or beast;
> In doubt his Mind or Body to prefer;
> Born but to die, and reas'ning but to err;

Chaos of Thought and Passion all confus'd,
Still by himself abus'd, or disabus'd;
Created half to rise, and half to fall,
Great lord of all things, yet a prey to all;
Sole judge of Truth, in endless error hurl'd;
The glory, jest and riddle of the world.

One last point about the Great Chain of Being must be emphasized in conclusion. Just as in its predecessor, the Ladder of Nature, the relationship between the different grades of being in this scale in no way implied genealogy or descent. The assumption that organic species were fixed and unchanging was even more fundamental than the idea of the Great Chain itself. The different species of living creature were so many aspects of the Creator's original plan—so much evidence (as John Ray put it) of 'the Wisdom of God, manifested in the Works of the Creation'. Ray was one of those Protestant scientists who searched the natural world for evidence of the divine master-plan, and his *Wisdom of God*—published in 1691—was, in effect, a scientific sermon. He computed the number of different species of terrestrial beasts and birds, fishes and insects, remarking that 'How much more perfect any *Genus* or Order of Beings is, so much more numerous are the *Species* contain'd under it'. In addition to some 150 species of beasts and serpents, and several hundred species each of birds and fishes (excluding shell-fish), there were, he estimated, around 10,000 species of insects, and probably 18,000 or more different species of plants.

> What can we infer from all this? If the number of Creatures be so exceeding great, how great, nay, immense, must needs be the Power and Wisdom of him who form'd them all! . . . As it argues and manifests more Skill by far in an Artificer, to be able to frame both *Clocks* and *Watches*, and *Pumps*, and *Mills*, and *Granadoes*, and *Rockets*, than he could display in making but one of those sorts of engines; so the Almighty discovers more of his Wisdom in forming such a vast multitude of different sorts of Creatures, and all with admirable and irreprovable Art, than if he had created but a few; for this declares the greatness and unbounded capacity of his Understanding.

For John Ray, the prime evidence of God's wisdom was thus the unbounded richness of the world; and, since this wisdom was 'manifested in the Works of the Creation', it seemed certain that the existing 'corporeal Creatures' were a permanent aspect of Nature, whose species had been established according to the divine plan at the Creation, *and maintained ever since.*

The Great Chain of Being, and the belief in Fixed Species, provided a stimulus both to field naturalists and to comparative anatomists. Some travelled to new regions of the Earth, in search of unknown animals and plants—like Linnaeus' pupil, Solander, who accompanied Captain Cook on his first circumnavigation. Others dissected as many species of animals as they could obtain—like Edward Tyson, who in 1699 described the anatomy of an infant chimpanzee from Angola, believing it to be the same as the orang-outang and therefore the missing link. In either case, one was adding further items to men's knowledge of the Great Chain of Being, and so helping to demonstrate the full wonder of the divine Creation.

The Eighteenth-Century Commonplaces

At first sight, a great gulf separated the Cartesians and Deists from the Protestant scientists and natural theologians. Yet those who interpreted the Order of Nature theologically, and those who refused to do so, shared a common starting-point. Both parties accepted this same Order of Nature as the fundamental datum, assuming that the material world in all its aspects conformed to certain fixed laws. In the minds of eighteenth-century scientists, indeed, the phrase 'laws of nature' was no dead metaphor: the analogy with civil law was taken seriously, scientific laws being construed as rules laid down by the Almighty for His inanimate Creation. Between them, scientists and theologians built up a static and unhistorical system of natural philosophy, in which fixed species of living creatures and solid unbreakable atoms alike conformed to the unchanging laws and specifications appropriate to their various kinds.

The short Biblical time-scale, which still bounded men's speculative imaginations, strengthened the hold of this static picture. (Although mediaeval scientists such as Albert the Great could and did debate whether organic species were absolutely fixed, seventeenth-century naturalists had no such freedom.) The static picture, in return, helped to defer the moment of truth, when the true antiquity of the cosmos had to be faced and the framework of the natural order recognized as the transient end-product of historical development. For the moment, no one had formed any real conception of Nature as developing through time. Only in the realm of human affairs, and to a lesser extent in geology, were men beginning to appreciate the depth and reality of historical change; and even in the social sciences strong intellectual pressures were distracting attention from those changes—encouraging there, too, a belief in 'fixed laws' of human nature and social structure. To complete the background to the historical revolutions in nineteenth-century natural science, we must consider next the ideas which, since 1600, had been shaping human history and social theory.

FURTHER READING AND REFERENCES

The starting point for the seventeenth-century revival of interest in natural history—in particular, geology and cosmology—is the subject of many recent books: notably,

E. M. W. Tillyard: *The Elizabethan World-Picture*
Marjorie Nicolson: *Mountain Gloom and Mountain Glory*
Basil Willey: *The Seventeenth-Century Background* and *The Eighteenth-Century Background*

On Descartes and his wider influence, consult the recent selection of his philosophical works translated by Peter Geach and G. E. M. Anscombe, with its introduction by Alexandre Koyré, and also the chapters on the Cartesians in

J. B. Bury: *The Idea of Progress*

The beginnings of historical speculation in cosmology, geology and palaeontology are well discussed in

John C. Greene: *The Death of Adam*
Francis C. Haber: *The Age of the World*
Forerunners of Darwin, 1745–1859: ed. Glass, Temkin and Strauss

For the background to the biological sciences, see

A. O. Lovejoy: *The Great Chain of Being*
C. E. Raven: *John Ray, Naturalist*
Loren Eiseley: *Darwin's Century.*

5

The Revival of Civil History

HUMAN affairs change more rapidly than the face of the Earth or the forms of living species. Less than six thousand years ago, the habits of city-dwelling and social life were restricted to a few river-valleys in Asia and the Middle East, and the arts of metal-working were in their infancy. We might, accordingly, expect men to have acquired a sense of temporal development first and foremost in the human sciences—recording and reflecting on the course of their own destinies—rather than by theorizing about the possibility of geological and zoological change.

To some extent this expectation is fulfilled. Around A.D. 1400, one detects the first recognition that social structures have changed significantly down the centuries. Yet a dozen obstacles delayed the moment when men in Western Europe could enquire once again, as the Ionians and Epicureans had done, into the origins of the family, society, language and the basic human crafts. The growth of the historical consciousness turned out to be a slow business, even in the field of human affairs. Before the nineteenth century, hardly anyone understood quite how deeply time leaves its mark on human affairs, and quite how far questions about historical development have to be pressed. Curiously enough, the final impulse came in part from the natural sciences: the expanding time-scale of geology and zoology giving historians a new freedom to manœuvre and speculate.

The Birth of Historical Criticism

In civil history, as in natural science, men's eyes were opened to the fact of historical development by a recurrent sequence of similar problems. These were all concerned with the concept of *authority*—the right of secular and ecclesiastical powers to impose their will and demand obedience. Church and State, King and Parliament, common law and popular will: at every stage, questions about the validity of political obligations led on to questions about their historical origins in the hope that the scope and limits of this authority might be discovered somewhere in the past.

This process began in the fifteenth century. Long before Luther's desperate renunciation of Papal jurisdiction, the Western Church was already struggling to reform itself from within. From 1431 on, the General Council of Basle had debated inconclusively the question, what secular powers the Papacy could legitimately exercise. For centuries the Church had claimed political authority over the successor-states of the Western Roman Empire, and the Pope had assumed the right to nominate a 'Holy Roman Emperor' to whom these temporal powers were delegated. This Papal claim to a temporal Empire had given rise to confusion between secular and sacred authority, and between spiritual and political allegiance. With his usual extraordinary perceptiveness, Cardinal Nicholas of Cusa saw where the heart of this problem lay, and argued that all political claims should be abandoned—the Church so returning to its original, spiritual mission.

Cusa's memorandum to the Council of Basle was one of the first examples of true historical criticism in the whole of European thought. As Nicholas pointed out, the foundation of the Pope's political claims was the *Donation of Constantine*—a document by which the Emperor Constantine the Great (fourth century A.D.) had allegedly divided the Roman Empire in two, transferred imperial authority over the Western Provinces to Bishop Sylvester of Rome and his successors, and reserved for himself only the Eastern Empire, governed from his newly-founded capital of 'Constantinople'. By the 1430s, the authenticity of this document had been accepted without question for more than five hundred years: in mediaeval political thought it had occupied a place like that of the Declaration of Independence in the American Constitution today.

Now Nicholas declared, quite baldly, that the *Donation* was a forgery. Who exactly had forged it, or when, or where, he did not presume to say: he was concerned only with the fundamental fact. A little knowledge of fourth-century Rome was enough to show that it could not be genuine. The terms in which it was drafted, the powers it purported to transfer, the assumptions it made about history and society, even its style and vocabulary: the document was full of *anachronisms*—forms of words, assumptions, implications and ideas quite out of harmony with its alleged origin in fourth-century Rome. Without using the modern term 'anachronism' to describe these incongruities, Nicholas' argument grasped one essential axiom of modern historical criticism—the organic unity of an age, by which genuine relics of former times can be sifted from fraudulent ones, and possible theories from inconceivable ones. A few years later his secretary at the Council, Lorenzo Valla, expanded Nicholas' historical argument into a full-scale treatise. In the meantime, Valla had left the Cardinal's service to join the court of King Alfonso I of Aragon, Sicily and Naples. Alfonso had his own particular motives for resisting the

Pope's secular claims, and welcoming any proof that the *Donation of Constantine* was spurious: his territories in Southern Italy were in a particularly vulnerable position. Valla's *Treatise* was accordingly a useful propaganda weapon as well as a model of historical analysis.

Scholars now believe that the *Donation* was counterfeited during the eighth century—possibly within a few years of 760, at a time when Pope Paul I was anxious to cut his ties with the iconoclastic authorities of the Byzantine Empire. Valla could still prove nothing positive about the origin of the forgery, but he pieced together a formidable case against its authenticity. He threw himself into this task with vigour and enthusiasm. The *Donation* relied on a legend according to which Pope Sylvester had cured Constantine of leprosy, and the Emperor in return had both named Christianity as the Established Church throughout his dominions and also, as an additional token of gratitude, surrendered to the Roman Pontiff imperial authority over 'the city of Rome, Italy and the western parts of the Apostolic See'. In page after page of textual criticism and heavy sarcasm, Valla demolished the forger's work, with all the murderous jollity of a dancing steamroller:

> O thou scoundrel, thou villain! the same history [the *Life of Sylvester*] which you allege as your evidence, says that for a long time none of senatorial rank was willing to accept the [new Christian] religion, and that Constantine solicited the poor with bribes to be baptised. And you say that within the first days, immediately, the Senate, the nobles, the satraps, as though already Christian, with the Caesar passed decrees for the honouring of the Roman Church! What! how do you want to have satraps come in here? Numskull, blockhead! Do the Caesars speak thus? Are Roman decrees usually drafted thus? Whoever heard of satraps [a Persian word, revived in eighth-century Rome] being mentioned in the councils of the Romans?

The dissection was merciless and exact. For instance, the *Donation* ostensibly conferred the title of 'consuls' on all members of the Roman clergy: Valla made fun of this claim too—

> But how can the clergy become consuls? The Latin clergy have denied themselves matrimony; and will they become consuls, make a levy of troops, and betake themselves to the provinces allotted them [as consuls] with legions and auxiliaries? And are there to be not two, as was customary; but the hundreds and thousands of attendants who serve the Roman Church, are they to be honoured [like fourth-century consuls] with the rank of generals?

Looked at in an historical perspective, the suggestion was laughable. The traditional consulate—with two consuls appointed to serve in each year—lasted unchanged until the German invasions in the fifth century A.D.: only much later was the *word* 'consul' revived to denote a particular grade in the social order. The use of the word in the *Donation* was as though the Anglican clergy refused to pay their taxes in the twentieth century, and produced in justification from the files at Canterbury a decree ostensibly signed by King Henry VIII declaring that he 'excused them from income tax in perpetuity'!

From 1450 on, the opposition between the all-embracing claims of the Church and the new demands of a more secular age continued and sharpened, providing the driving-force behind the reviving interest in civil history. Throughout the sixteenth century, problems about legal authority continued to act as a spur to historical enquiry. In France, for instance, 'legal humanists' such as Cujas and Bodin faced some fundamental questions about the contemporary French system of law, which was based on codes inherited from the Romans: what perplexed them was not the validity or authenticity of Roman law, but its exact relevance to the changed conditions of life of their own country and century. Such humanists might dismiss the theories of Epicurus and Lucretius as atheistical, but they were none the less secular scholars rather than doctors of the Church. Condemning the limitations of scholastic education, they gradually helped to tip the scale in favour of a much wider educational system—one which was wider in scope, through its emphasis on the newly-recovered classics, and also more widely available, through the foundation of the new grammar-schools. (By 1500 these were attracting more charitable gifts and bequests even than the Church itself.) Once literature began to be valued alongside divinity, the ancient masters of historical narrative and portraiture were soon treated with as much intellectual respect as Eusebius and Jerome. So Herodotos, Thucydides and Plutarch began once again to exert a significant influence on European thought.

Decay or Progress?

Despite all their vigour and originality, however, Renaissance historians no more evaded the influence of sixteenth-century 'common sense' than did their colleagues in the natural sciences. In an age when it was universally assumed that 'the World grows near its end', both historical and scientific ideas had to accommodate themselves to the coming Apocalypse. So it is no surprise to find the deep pessimism we have already observed in theology and literature reappearing around 1600 in historical writings also: more surprising, perhaps, are the first breaths of wind from a more optimistic direction—the first suggestions that the social and political fortunes of this world may, after all, be capable of future improvement. These opposed attitudes to history are exemplified by two English historians of the early seventeenth century. Both men were distinguished figures, looking back on long careers in public service, who late in life found themselves disgraced. Sir Walter Ralegh's *History of the World* was begun around 1604 when he was a prisoner in the Tower of London, under sentence of death for high treason: Francis Bacon wrote his *History of Henry VII* in 1621, when he too was experiencing the bitterness of political downfall.

In every other respect, the two men's histories are in complete contrast. Despite Ralegh's wide reading and experience of the world, his approach was that of a mediaeval. Like Augustine more than a thousand years before, he aimed to demonstrate the action of divine Providence through historical events. The framework of human history had been set down, once for all, in the Biblical story of Man's Fall and God's Mercy. Holy Scripture had for him an authority as unquestionable as the Koran has for an orthodox Muslim and, for the truth about the events recorded in the Bible, the historian need not look beyond it. It would be a waste of time to search 'as it were by candlelight, in the uncertain fragments of lost authors', for truths 'which we might have found by daylight, had we adhered only to the Scriptures'.

The historian's task was to fill out this scriptural framework, collecting from the classical and modern periods narratives which could 'teach by examples of times past such wisdom as may guide our actions'. The real significance of historical events—their 'true and first causes'—lay in their relevance to God's Providence alone. The changes and chances of political fortune were only 'second causes'—the 'instruments, conduits and pipes, which carry and displace what they have received from the head and fountain' of God. As for the chronology of his world-history: like Jerome and Luther before him, Ralegh found this in the Biblical narrative. (His own date for the Creation was 4032 B.C., on which basis the world's approximate life-span of six thousand years would be due for completion in A.D. 1968.)

So far as the presuppositions of the time permitted, Francis Bacon's ideas about history went to the other extreme. In his actual narrative, he paid little more than lip-service to divine Providence, and in his discussions of historical method he allowed it no great place. The aim of history was not to provide edifying tales for purposes of moral instruction: on the contrary, its chief value was as the basis for a future science of Man. All those 'second causes', which had been of minor importance to Ralegh, became for Bacon the chief object of historical interest:

> Above all things (for this is the ornament and life of Civil History), I wish events to be coupled with their causes. I mean that an account should be given of the characters of the several regions and peoples; their natural dispositions etc. etc.... Now all this I would have handled in a historical way, not wasting time, after the manner of critics, in praise and blame, but simply narrating the fact historically, with but slight admixture of private judgement.

In this way, history would serve the general aim of all science and learning —'that human life be endowed with new discoveries and powers'.

Bacon did not share the humanists' preference for the writers of classical Greece and Rome—*prejudice*, he would have called it:

> Men have been kept back as by a kind of enchantment from progress in the sciences by reverence for antiquity, by the authority of men accounted great in philosophy, and then by general consent.

This reverence for antiquity originated largely, he claimed, from sheer confusion of thought: the antiquity (i.e. age) of the documents reaching us from classical times had been taken as guaranteeing the antiquity (i.e. venerability) of the ideas expressed in them. In fact, he retorted, the humanists' perspective was the reverse of the true one—the more antiquated the ideas, the more youthful a stage in human thought they represented. *Antiquitas seculi juventus mundi:* the period the humanists so much revered had been the childhood of the world.

So Bacon became one of the first true modern historians. Instead of confining himself to sacred history, or to Greco-Roman times, he was prepared to treat the men and events of his own age as significant in their own right. If anything, he argued, the men of modern times had built on the achievements of the ancients and so surpassed them. Printing, gunpowder and the compass 'have changed the appearance and state of the whole world . . . no empire, sect, or star appears to have exercised a greater power of influence on human affairs than these mechanical discoveries'. Nor could one suppose that the age of discovery was over— 'this proficience in navigation and discoveries may plant also an expectation of the further proficience and augmentation of all sciences', and by 'prudent interrogation' of Nature human power and understanding could be greatly increased.

Bacon's intellectual optimism broke completely with the mediaeval tradition. Only a short part of the world's life-span might remain, but there was no reason why these last years should not also be the best. 'In proportion as the errors which have been committed impeded the past, so do they afford reason to hope for the future', since they put us in a position to improve human welfare, in a quite worldly sense. Bacon did not mince his words. Whereas the Puritans had applauded the severe Augustinianism of Sir Walter Ralegh's *History of the World*, Bacon's *Advancement of Learning* and *New Atlantis* proclaimed, frankly and openly, that comfort and happiness in this world were worth having for their own sakes and that men were justified in extending their command over Nature as a way of achieving these ends. This new secular tone was as different from the dark prophecies of Luther and Calvin as it was from the other-worldly resignation of Augustine. Protestantism was beginning to undergo a subtle but fundamental change: the idea of Grace was giving way to the idea of Self-Help.

How much longer did Francis Bacon expect the world to last? Though he spoke of his own times as the Old Age of the Earth, he gave no sign of regarding the final Conflagration as imminent. His attitude to the Death of the World was the attitude that most men adopt towards their individual deaths: he recognized it as inevitable, but did not allow it to prey on his mind or stand in the way of useful occupation so long as life lasted. Thus the new worldliness grew up within the boundaries of the old timescale. Mankind, in Bacon's view, had not become frail, dissipated or decrepit, but was rather in a ripe old age, and the important thing was to make the most of these final years. The future might be better, but it would not necessarily be any longer.

In the years that followed, however, one can see the threat of cosmic extinction perceptibly lifting. George Hakewill, in his *Apologie or Declaration of the Power and Providence of God* (1627), flatly denied that the Fall of Man had infected the whole of Nature. He did not claim that the world was actually eternal—that would be the Aristotelian heresy—but its end would come about solely by a supernatural act of the Deity, and we had no reason to suppose that this was imminent. As he said:

> It is agreed upon all sides by Divines that at least two signs forerunning the world's end remain unaccomplished—the subversion of Rome and the conversion of the Jews. And when they shall be accomplished God only knows, as yet in man's judgement there being little appearance of the one or the other.

In practical affairs, indeed, we should ignore the prospect of extinction:

> The opinion of the world's universal decay quails the hopes and blunts the edge of men's endeavours. . . . Let not then the vain shadows of the world's fatal decay keep us either from looking backward to the imitation of our noble predecessors or forward in providing for posterity, but as our predecessors worthily provided for us, so

> let our posterity bless us in providing for them, it being still as uncertain to us what generations are still to ensue, as it was to our predecessors in their ages.

By the standards of the time, this was radical stuff; but forty years later the threat had largely vanished. The founders of the Royal Society worked confidently on behalf of posterity—confident, at least, that there would be a posterity to enjoy the fruits of their labours. As Joseph Glanvill put it: 'We must seek and gather, observe and examine, and lay up in bank for the ages that come after.' And by the time Andrew Marvell wrote the ode *To His Coy Mistress*, he could juggle with spans of time which the men of his grandfather's day never contemplated. The shadow of extinction now lay only over men's individual lives.

> Had we but world enough, and time,
> This coyness, lady, were no crime . . .
> . . . I would
> Love you ten years before the Flood,
> And you should, if you please, refuse
> Till the conversion of the Jews;
> My vegetable love should grow
> Vaster than empires and more slow;
> An hundred years should go to praise
> Thine eyes, and on thy forehead gaze;
> Two hundred to adore each breast,
> But thirty thousand to the rest;
> An age at least to every part,
> And the last age should show your heart . . .
>
> But at my back I always hear
> Time's wingèd chariot hurrying near,
> And yonder all before us lie
> Deserts of vast eternity.

The Customs of the Ages

Meanwhile, the rapidity of social change was forcing historical scholarship along the secular channels that Bacon had indicated. In England, for example, there was a continuous tussle between Parliament and the Crown, with both sides appealing to history in justification of their political claims. By an accepted principle of common law, any custom or institution which was 'immemorial' (i.e. which could be shown to antedate all surviving records) required no formal sanction or warrant. So Parliamentarians attempted to trace the history of the House of Commons back continuously to the Anglo-Saxon 'folk-moots' before the Norman Conquest; while, in return, Royalists insisted that Parliament was a novel institution, dependent for its legality on a royal writ of summons, and dating at the earliest to Henry III. Both parties in this debate had strong motives for studying the actual history of parliamentary powers and royal prerogatives: as a result, the early seventeenth century saw the first detailed analyses of the structure of mediaeval society—in particular, Sir Henry Spelman's study of 'feudalism'.

Just as the Crown had to defend its claims against the new challenge from the Commons, so too the ecclesiastical authorities were under fire from laymen. The main question in dispute was concerned with the traditional right to levy 'tithes'. The same scrupulous care that Spelman had devoted to feudalism and Parliament was shown by his contemporary, John Selden, in his *Historie of Tithes*—a book which many contemporaries read as an attack on the Church's right to tithes, even though Selden had carefully confined himself to the historical facts about this particular practice. The truth was that many of the stock arguments about tithes were weak or even fantastic, involving (for example) Biblical references to Melchizedek, and the existing situation had evolved in a much more complicated way than anyone had hitherto realized. No single argument could justify all the rights claimed by the ecclesiastical authorities, and any particular claim would have in future to be defended on its own merits.

The acid of historical criticism did not act only on the Church's secular claims. It also changed men's attitudes to the spiritual authority of Church and scripture; so that, by 1700, an educated man could no longer revere the transmitted words of the Bible as unquestioningly as Sir Walter Ralegh had done a hundred years earlier. There were two reasons for this change. In the first place, it had become doubtful what exactly were the authentic words of the Holy Bible: when humanist scholars such as Erasmus and Scaliger compared the Latin Vulgate of the mediaeval church with Greek manuscripts of undoubted antiquity, they found unsuspected disagreements between the different versions. So a running fight began between theologians and Biblical scholars, partisan motives driving them deeper and deeper into grammar, philology and ecclesiastical history; and at the end of the seventeenth century this debate was still going on, when Father Simon's *Critical History of the Old Testament* launched the Catholic party unwillingly and apprehensively on to the stormy waters of 'higher criticism'.

The other factor which injected an historical element into the religious argument sprang from the very kernel of the Reformation debate. Luther had taken over Lorenzo Valla's *Treatise on the Donation of Constantine* as a weapon against Papal supremacy, and from that moment the rival claims of the Catholic, Lutheran, Calvinist and Anglican confessions were open to historical attack. The Protestant claims were necessarily based on appeals to history, and so could only be as secure as historical research made them. For, once men appealed behind the received tradition of the Catholic Church to some supposedly purer and more authentic doctrine, they were forced to establish by acceptable historical arguments the precise respects in which Catholicism had deviated from the original teachings of Christ. So theology, as well as law and politics, was thrust into the arms of history—compelled to reconstruct the truth about Christian doctrine out of evidence recovered (in Bacon's vivid image) 'like the spars of a shipwreck . . . from the deluge of time'.

Whatever their origins, the outcome of these historical researches was the same: men were compelled to acknowledge the variability of social customs, and the corresponding relativity of laws and obligations. A code of law framed in Imperial Rome was no longer at home in sixteenth-century France, and there was the same need to consider all institutions in their proper contexts. The point was well stated by Spelman, in the introduction to his essay on the history of Parliament:

> When States are departed from their original Constitution, and that original by tract of time worn out of memory; the succeeding Ages viewing what is past by the present, conceive the former to have been like to that they live in, and framing thereupon erroneous propositions, do likewise make thereon erroneous inferences and Conclusions.

Each age (he implied) has its own character and customs; and these differences have to be taken into account before one can safely compare their achievements, institutions or legal systems.

Like Lavoisier in eighteenth-century chemistry, Spelman began his greatest work as a systematic analysis of the vocabulary of his subject, but what began as a philological enquiry ended in both cases as a full-dress reconsideration of principles. Having planned to compile a dictionary of obsolete legal jargon, Spelman found himself compelled to analyse the whole structure of mediaeval society, and to expound the different technical terms in terms of this analysis. Philological studies thus became a powerful instrument of historical study, and it was no accident that the most learned English historian of the age, John Selden, was also a formidable linguist.

It was Selden himself who saw most clearly the principles of method which must goven historical research in this new era. The historian should build up a consistent and coherent picture of the life, society, law, religion and customs of every period and country. The dating and authenticity of any historical relic should then be judged by seeing how it fitted into the established picture of the place and time to which it was attributed. Selden thus extended and described the method that Lorenzo Valla had used to expose the spurious *Donation of Constantine*. Historical questions were concerned with 'synchronism'—this word primarily meaning 'simple chronological consistency'. The word 'anachronism', its opposite, likewise meant in the first place a simple misdating, or chronological inconsistency—like dating a man's signature to a time after his death. Yet Selden's own historical arguments were also concerned with a wider kind of consistency and inconsistency—with the 'erroneous inferences and conclusions' that arise from those inexact historical analogies which Spelman had criticized. So, arising out of the idea of historical *periods* came the idea of historical *incongruities*—what Coleridge christened 'practical anachronisms'—the sense in which the word is used today.

Human Nature and Social Change

The discovery that every society and period has its own special character did not, however, lead on at once to a recognition of *historical development*. From our point of view, this appears the obvious next step: having admitted changes between one period and the next, surely the next problem is, how does each phase develop out of the preceding one? Yet to consider historical periods as successive phases in a continuing sequence is the logical next step, only if one has already begun to see things in an 'evolutionary' perspective, and this point of view did not come naturally to seventeenth-century thinkers. In civil history, as in natural philosophy, there was an intermediate stage before our own, during which the course of history was conceived not as a long-term evolution, but rather as a sequence of variations on a theme. This intermediate point of view established itself around 1700, and provides a nice illustration of 'synchronism' in the history of thought. For the same general factors that inclined natural philosophers to believe in a fixed and static Order of Nature had an influence also on historians and social scientists. Having rejected the doctrine of universal decay, they replaced it by the theory that society operated in accordance with fixed causal laws, having their roots in human nature. At the time, that was as far as optimism could go.

The debate about the constancy or decay of human nature had originated as a by-product of humanism. Renaissance scholars fell so much in love with the classics of antiquity that many of them despaired of the human race ever again reaching the same heights. They lavished on

Sophocles and Cicero, Thucydides and Livy, Plato and Horace, the same devoted admiration that mediaeval philosophers had conceived for Aristotle and, for a while, this touching passion helped to confirm the theological belief that humanity was corrupted and in decay. So, when a less despairing generation arose, they found themselves at odds with the humanists as much as with the scholastics; and the literary battles which ensued came to be known as the war between the Ancients and the Moderns.

These battles rumbled on intermittently for more than a hundred years, petering out in the early 1700s after the final explosion of Jonathan Swift's *Battle of the Books*. The most noisy engagement took place in France during the 1680s, the chief partisans being Boileau for the Ancients and Perrault for the Moderns. It was the age of Louis XIV. Responding to the general buoyancy of the time, Perrault argued that the Moderns had nothing to fear from a comparison with Plato and Homer: on the contrary, the present age was in some respects more enlightened than the earlier. Yet, for all his pride in the Moderns, even Perrault did not claim that the men of his time were actually *more* talented than their classical predecessors. He was content to argue only human powers were *constant.*

À former les esprits comme à former les corps
La Nature en tout temps fait les mesmes efforts;
Son être est immuable, et cette force aisée
Dont elle produit tout n'est point épuisée; . . .
De cette mesme main les forces infinies
Produisent en tout temps de semblables génies.

Shaping minds or shaping bodies
Nature exerts herself alike in every age;
Being essentially unchanging, the effortless power
By which She creates all things cannot be used up; . . .
The limitless powers of this same hand
Create comparable spirits in every age.

This idea chimed in happily with the fashion for the Cartesian system. When Descartes stated his axiom that the natural powers in the physical world had not changed since the Creation, this denial had reverberations outside physics. His contemporaries seized on the principle as offering the foundation for a new view of human affairs also; for why should the mental and moral capabilities of the human race have lessened with time, any more than the 'quantity of motion' in the physical world?

This was certainly the position of that good Cartesian, Bernard de

Fontenelle. In 1683, he published his *Dialogues of the Dead*, a series of imaginary conversations which pitted Harvey against Erasistratos, Montezuma against Cortez, and Socrates against Montaigne. By a nice ironical twist Montaigne, the modern French sceptic, was made to argue that the great men of Antiquity had no equals today, while Socrates, the ancient philosopher, defended the Moderns. ('Elsewhere,' Socrates argued, 'Nature has lost none of her powers: the trees of the present day are as tall as they ever were, so why should men alone have degenerated?') Returning to the subject a few years later, Fontenelle made the connection between human nature and Cartesian physics even more explicit. Mind and matter are—he allowed—distinct: none the less, our mental and moral capacities depend on the bodies and brains with which Nature has equipped us. Now, the axioms of Descartes' system apply equally in dynamics, astronomy and physiology, and all the evidence is that plants, animals and men still have today the same physiological structures as those they had two thousand years ago. We must therefore conclude that mentally also they are potentially the equals of the Ancients. If the achievements of different ages and societies vary in character and merit, the reason must lie, not in man's innate constitution, but in external factors—climate and commerce, human experience and prejudices, social conditions and cultural interactions.

By the early eighteenth century, the war between the Ancients and the Moderns was over. The constancy of human nature was generally conceded, and the problem now became to explain the differing characters of the various ages and nations. Conservative scholars still followed Bossuet's *Discourse on Universal History*, which interpreted the whole of human history as the working-out of divine Providence. But this was consciously a rearguard action against libertines and free-thinkers: by the 1740s, the new mechanical philosophies had the upper hand, and Montesquieu was able to publish his *Esprit des Lois*, a general analysis of the differences between societies, in which the workings of providence were re-interpreted as the outcome of certain 'general causes'.

Society, in Montesquieu's view, conformed to fixed laws, parallel to those of the physical world. The ups-and-downs of politics are neither rewards or punishments, nor effects of chance: they are, rather, the products of those 'general causes, moral or physical, which operate in every [political system], raise it, maintain it, or overthrow it'. These same general causes, operating in a variety of situations, explain also why, at any given time and place, men take up those attitudes to politics and social life which together form the 'common social attitude' (*esprit générale*) of that period and nation. As for their legal codes and procedures, which at first glance differ so much from age to age: these differences too represent, not contradictions or inconsistencies, but 'necessary relations deriving from the nature of things'—naturally-varying responses to differing social situations. The totality of 'social causes' shapes the institutions of every society into an organic whole, just as the totality of 'physiological causes' does in the case of the human body.

What kinds of social factors did Montesquieu include among his 'general causes'? They were something of a mixed bag. He placed a great deal of importance on geography and climate:

> Cold air shrinks the extremities of the exterior fibres of the body; this increases their elasticity, and encourages the return of the blood from the extremities to the heart. It reduces the length of these same fibres; thus it increases their force still more. Warm air, on the contrary, relaxes the extremities of the fibres, and lengthens them; so it reduces their force and energy. Thus one has greater vigour in cold climates. . . . In cold countries, one will have little feeling for the pleasures; more in temperate countries; in warm countries, it will be extreme. As climates are distinguished by degrees of latitude, they could be distinguished (so to speak) by degrees of sensitivity. I have seen the opera in England and in Italy; the same pieces done by the same actors; but the same music produces such different effects on the two nations—the one so calm, the other so transported—that it seemed inconceivable.
>
> [Elsewhere] the barrenness of their lands makes men industrious, sober, hardened to toil, courageous, fit for war; they are obliged to procure for themselves what the soil denies them. Fertility gives, with an easy life, softness and a certain desire for self-preservation. It has been remarked that German troops levied in regions where the peasants are rich (such as Saxony) are not as good as the others.

In Mediterranean countries, Montesquieu argued, there was a natural inequality of the sexes, since the women were ready for marriage very young, before their reasons were well formed:

> When beauty called for dominion, reason refuses it; when reason could obtain it, beauty has gone. So the women have to be in a dependent state: for reason cannot procure them, in their age, the dominion beauty had not given them in youth.

Where religion did not forbid it, the natural consequence of this was polygamy. In temperate lands, on the other hand, there was a greater equality of the sexes and monogamy was the general rule: in excessively cold countries, indeed, strong drink produced intemperance among the males, and the women more easily obtained the upper hand.

Alongside geography and climate, Montesquieu's analysis of 'general causes' referred also to social traditions. Inherited manners and customs could leave an enduring mark on law and social structure. Thus, in some Indian castes, the women considered it dishonourable to learn to read: that was the job of the slaves who sang hymns in the temples. So rules established first by pride acquired the force of laws. Now at this point one might argue that Montesquieu was abandoning his own axioms. Customs and prejudices should in principle have been treated as 'social effects', not as 'causes'. By bringing traditions into his list of general causes, he was implicitly admitting that the Social Order represented, not an equilibrium between a fixed set of causes, but a system of relations developing through time. Still, though his list of social causes may have been confused, the *Esprit des Lois* was a landmark in sociological thought—and the first serious treatise on social science: the questions he put into circulation have kept a place in our ideas ever since.

Some of Montesquieu's modern admirers have expressed their disappointment that he formed no real conception of progress or development, and had no feeling for Time itself as a factor working to modify the patterns of social life. Ancients, mediaevals and moderns; Mediterraneans, Asiatics and North Europeans—to him, all men were one, and later thinkers have accused him of wilfully muddling together societies and periods which are not comparable at all, since they represent different stages in social evolution. If only Montesquieu had grasped the Idea of Progress, he might have recognized that different societies are not just an incoherent collection of human populations, reacting in their respective ways to variations in environment and tradition, but rather an orderly sequence linked by growth and genealogy: so many products of a single, all-embracing historical process—so many milestones along the path of human development.

This reaction is irrelevant—the response of nineteenth-century minds to a man who saw society through eighteenth-century eyes. There are very good reasons why Montesquieu should have compared and contrasted societies of all kinds and from all periods. If he 'abstracted social phenomena from their relations in time', this was not (as Bury complains) 'a fundamental defect' of his method. Rather, it was the essential first step towards social science as we now know it, and it was one which reflected the convictions of his age—notably the belief, which he shared with Fontenelle and Descartes, in the stability of human nature. If human beings have been alike in every race and age, then surely we may set the most disparate societies alongside one another, and look for external factors to explain their variety.

Montesquieu's social science was, accordingly, a mature product of the new philosophy: the very aspect of his thought which modern critics have attacked marks him as a true Cartesian. To the objection that his cosmology rested on arbitrary suppositions, Descartes had replied—

> It is hardly possible to make a supposition that does not allow of our inferring the same effects (perhaps with more labour) according to the same laws of nature. For according to these, matter must successively assume all the forms it admits of; and if we consider these forms in order, we can at last come to that which is found in this universe.

The phenomena of the natural world were so many permutations of a fixed theme, like successive changes in a peal of bells. Montesquieu's attitude to society was the same. Men are what they are, the world is as it is: so human history, too, represents a Theme with Variations, in which the different environmental factors combine in parallel and in succession, to work themselves out through society. We do not look to the *Goldberg Variations* for the internal development-through-time characteristic of a movement in sonata-form; nor is this absence of a temporal logic a reason for criticizing Bach. Montesquieu is entitled to a similar defence. He had grasped, we may say, the fundamental harmony and counterpoint exemplified in human society. The dynamics of social growth and change, like 'development' in music, could fairly be left for a later generation.

The Myth of the Social Contract

So, in one important respect, Montesquieu's programme for social science was unhistorical. He did not, of course, share Descartes' low opinion of 'histories and fables': one could scarcely investigate the general causes of social phenomena at all, unless one accepted some historical reports as true, faithful and interesting. But it was only the *data* of Montesquieu's social science that were historical in character: his *interpretations*, on the other hand, were mechanical—the structure of any society being the equilibrium of the 'social forces' acting upon it. In short, his was a theory of social statics, rather than one of social development.

Most social theory in the early eighteenth century shared this unhistorical character. The mediaeval doctrine of a 'natural law', reflecting the fixed nature of Man, survived till the 1760s, to serve as the central axiom of Blackstone's *Commentaries on the Laws of England*. Behind all the apparent variations and inconsistencies of judicial history, Blackstone claimed, the 'wisdom of our ancient law' had 'determined nothing in vain'; for the 'general spirit of laws and principles of universal jurisprudence' had remained fixed and unchanged throughout. At this level of sublimity, all the contrasts between one society and another, one period and another, were illusory: the 'general principles of universal jurisprudence' applied with equal force to men in every epoch. But the most influential product of this static attitude towards man and society was probably the doctrine of the 'social contract', by which philosophers explained the basis of political obligation.

The social contract theory is a late, yet typical, example of a 'justificatory myth'. We may compare the appeal to an 'original social contract', as a way of demonstrating the reasonableness of the existing Social Order, with Plato's appeal to his myth of the Craftsman to explain the rationality in the Order of Nature. Plato accepted the natural order as fixed and unchanging: so he could not consistently give a *historical* account of its origin, in which the present rational order developed out of a primaeval chaos gradually, through an intermediate sequence of less rational and orderly states. The Order of Nature could have originated in time in only one way—all at once, and once for all. Nor did it matter *when* this 'origin' was supposed to have occurred, provided only that it made the present Order of Nature appear rational. The idea of an 'original contract' played a similar part in European social thought. The problem was to explain why human beings born into a pre-existing social order must obey the established political authorities, and accept a whole code of laws which they had no hand in framing. Once the religious basis of political obligation was called in question, this problem became acute and urgent; and the social contract theory offered an attractive solution.

In the earliest days of human life, it ran, men lived harsh and solitary lives, every man's hand being turned against every other's, and the family being the largest social unit. In this 'state of nature', the life of Man was (in Hobbes' memorable phrase) 'nasty, brutish and short', and men banded together into a social order as a way of escaping from an intolerable anarchy. Society accordingly originated by men's voluntary choice, and their collective agreement to renounce violence and mayhem—the 'social contract'—bound both them and their descendants to accept a common system of obedience. On this account, present-day political obligations were justified by appeal to the continuing validity of the 'original contract'.

As to its actual terms, monarchists and democrats held different views. Some declared that the founder-members of the Social Order had pooled their resources for collective defence and handed over absolute sovereignty to a single 'monarch', in return for his maintaining their common security. According to others, the original contract was made between the original citizens individually, its effect being to create a constitution protecting all the citizens and their interests equally.

Either way, the social contract theory was vulnerable to the criticism

Aristotle had levelled against the *Timaeus*. The existence of binding political obligations *now* was a fact which nobody seriously questioned: the problem was simply to show that this fact was intelligible and reasonable, and that was just what the theory failed to do. For it made present-day obligations, about which there was little doubt, depend for their validity on a hypothetical event at some indeterminate moment in the past—an event whose historical authenticity was certainly open to question. And what if historians proved that there had never in fact been any such original contract? Would all political associations immediately be dissolved? Could the *non*-occurrence of the hypothetical contract imply the *non*-bindingness of political obligations? Secure in his own theory of human nature, David Hume insisted that the 'social contract' was quite irrelevant to all questions about the present social order. The only sure basis for present-day political obligations must lie in present-day human nature: the alleged social contract was a mere fiction.

Yet the fundamental weakness of the social contract theory lay deeper than Hume could suspect. It took for granted a contrast between two completely opposed conditions of life: either men were living in a 'state of nature' lacking all social organization, or else there already existed a fully-fledged 'social order'. The difference between the two conditions was absolute, and the supposed contract by which men passed from one condition to the other was a transition as abrupt and complete as the establishment of the Order of Nature by Plato's Craftsman. Indeed, few philosophers seem actually to have regarded the 'state of nature' as a dateable phase of human prehistory at all. There was a reason for this. When Hobbes characterized pre-social life as 'brutish', his choice of adjective went to the heart of the matter: the absolute contrast between anarchy and society reflected another deeper contrast—that between *human beings*, who had reason, and *animals*, which had not. Lacking reason—the common view implied—men would have lived like animals and suffered accordingly. Possessing it, they naturally saw the virtues of living peaceably in society. And, so long as men and animals preserved their fixed places in the Great Chain of Being, this contrast would presumably be a permanent aspect of Nature.

In this way, the social contract theory and the doctrine of fixed species dovetailed in men's minds, and the few thinkers who questioned the uniqueness of Man's social attributes also felt bound to question his unique zoological status. Monboddo, for example, believed that orang-outangs could be trained to play a part in human society, just because there were no significant zoological differences between the 'wild man of the woods' and civilized men in cities (see cartoon, plate 8. Jean-Jacques Rousseau, similarly, buttressed his political theories by classing Man and the higher apes together as one zoological species: he stood the eighteenth-century argument on its head, by declaring that the 'State of Nature' was not after all so nasty and violent, and foreshadowed the romantic taste for the 'natural life' of the 'noble savage', free from the shackles of organized society. But these men were exceptions. For most of their contemporaries, the difference between rational 'human nature' and non-rational 'animal nature' was an axiom; and the contrast between animal gregariousness and human society remained absolute. So the Epicurean view, that human society was the end-product of a protracted historical process, during which the nastiness of animal life was left behind only gradually and painfully—the social order representing a continuation with refinements of an earlier animal order, rather than its sheer opposite—was still denounced almost universally as 'reducing Man to the level of the Beasts'.

Intimations of Progress

By the middle of the eighteenth century, then, the pendulum had swung a long way from the commonplaces of 150 years before. Instead of assuming that Nature and Man alike were corrupted and in their last days, many people were now equally convinced that human capacities and propensities were the same in all ages, being just one aspect of the permanent Order of Nature. Another century was to pass before the framework of men's ideas would be as generally 'progressive' as the thought of 1600 had been 'regressive', and that of 1725 'static'. Yet it might be misleading to present a snapshot of mid-eighteenth-century thought frozen in this position. Having got that far, the pendulum had only reached its mid-point, and it had farther to swing; by the third quarter of the eighteenth century, the considerations which were eventually to carry men farther can already be recognized in the background. To begin with, the new-found historical attitudes, crystallized in Montesquieu's concept of the *esprit générale* characteristic of any age or society, were having their effect throughout all of Western life and thought: when they reached into technology, science and linguistics, it became clear that, in these spheres at least, human achievements had not just changed disjointedly from period to period—they had progressed cumulatively down the ages. By the end of the century, historians of science and engineering were following up Bacon's hint that the discoveries of printing, gunpowder and the compass could themselves be surpassed by further discoveries: scholars, too, were embarking on those comparative studies of the 'Indo-European' languages which revealed their common ancestry and genealogical connections, and so (as Dr. Johnson put it) 'the pedigree of nations'.

One can recognize how universal the sense of 'period' had by this time become if one notices how it affected activities quite remote from sociology and jurisprudence. Consider, for instance, styles in painting. Looking at mediaeval paintings of supposedly antique scenes (e.g. those by the Flemish Primitives) one finds the characters represented, as a matter of course, in mediaeval costume. A fourteenth-century painter did not trouble to adapt the clothing of his figures to the supposed dates of the events portrayed: he saw no reason to do so. After all, clothes were clothes. Solomon, Holophernes, Augustine of Hippo, or a living merchant—why should he paint them in anything but clothes of kinds familiar to fourteenth-century men? The Renaissance passion for antiquity equally had its effect on painting. During the seventeenth century, painters developed a taste for showing their contemporaries in imitation togas or Roman armour, and in the mid-eighteenth century Sir Joshua Reynolds was still teaching that a portrait-painter could convey nobility only by representing his subject in classical costume. But the next generation of artists rebelled. A new scruple began to affect them: the carefully-draped togas of conventional portraiture struck them, not as noble, but as incongruous and ridiculous. For the first time, they felt it essential to match clothes to subjects, and to avoid the anachronisms involved in earlier styles. At the same time, the sense of period was making itself felt on the stage also. Traditionally, productions of Shakespeare had been dressed out of one and the same wardrobe, regardless of period: nobody had thought it incongruous to present *Julius Caesar* in seventeenth-century dress. Now, however, a new and exciting alternative suggested itself. For the first time, theatrical producers made it a matter of deliberate policy to stage Shakespeare's ancient-history plays in the costumes of the original period. At the time, this move had all the force and shock of an *avant-garde* gesture. Men experienced the same excitement at being shown Julius Caesar dressed in the supposed costume of his own time that theatregoers in the 1930s experienced when—reversing the move—they were shown Julius Caesar in the clothing of our own time: namely in Fascist uniform.

Meanwhile, a few radical philosophers like Voltaire and Turgot were beginning to formulate the ideals of 'social progress' which were to play so large a part in later political thought. The conditions of human life were not inalterable, they argued, still less in irredeemable decay: on the contrary, the spread of education together with an intelligent reform of the social order would enable us to improve them, and even to perfect them. Taking this doctrine as their starting-point, men such as Helvetius and Condorcet conceived the vision of society which led them to pin such hopes on the outcome of the French Revolution. Yet this first conception of a progressive society was still only a semi-historical one. It operated within the accepted view of human nature and society as governed by fixed and definite laws, and did nothing to change that view. To these philosophers of the 'Enlightenment', there was no fundamental mystery about the political future: the progress they had in mind was always progress within human society as Montesquieu had analysed it. The basic problem of politics sprang directly from Man's ambiguous status, 'half-angel and half beast'. Given that men were pulled in two directions, by reason and passion, the task was to order social institutions by legislation in such a way that the rational half of human nature could realize itself at the expense of the brutal and superstitious half. If so much of earlier history had been a story of misery and oppression, that was because hitherto passion (the greed of tyrants) and superstition (the obscurantism of clerics) had stood in the way of education and reform. Social progress was thus to come about, not by changing the fundamental character of man and society, but by understanding it, and so exploiting it. Once the 'laws' of human nature, the 'mechanisms' of social interaction, and the 'forces' responsible for political change had been grasped—all these notions being borrowed more or less consciously from the physics of Descartes and Newton—then Reason might rule, and the Ideal Commonwealth of men's dreams could at last be brought into existence.

This was the fairy-tale politics of the *Magic Flute*, rather than that of real life. The over-simplifications of Montesquieu's analysis made his successors over-optimistic. Having taken the fundamental powers of human nature as a fixed and inalterable starting-point, they underestimated the creative possibilities of history, and the difficulty of accelerating the processes of historical change by legislative action. As with the Russian Revolution in the twentieth century, many of the intellectuals who at first welcomed the French Revolution as an instrument for human progress were, like Condorcet, disillusioned by the aftermath. Yet, despite all its weaknesses, this naive view of politics as 'applied social physics' was intellectually a productive one. Under its influence, for example, Adam Smith and the Scottish historical philosophers laid the foundations for economic theory as we know it today. The nature of money, the methods of production, distribution and exchange, the division of labour: as a result of the eighteenth-century debate about human nature, these topics became matters for serious theoretical discussion. And, though in retrospect we may question the physical analogies—the 'laws', 'forces' and 'mechanisms'—in terms of which the discussion was so often conducted, the resulting tradition of systematic enquiry into economics and social institutions has been a very positive gain.

FURTHER READING AND REFERENCES

For the general topics of this chapter we are particularly indebted to

F. Smith Fussner: *The Historical Revolution*
R. G. Collingwood: *The Idea of History*

C. B. Coleman prepared the modern edition of Lorenzo Valla's *Treatise on the Donation of Constantine*. For sixteenth-century pessimism, see Marjorie Nicolson and E. M. W. Tillyard, as before. Particular aspects of seventeenth-century historical thought are carried farther in

R. F. Jones: *Ancients and Moderns*
J. B. Bury: *The Idea of Progress*
B. Farrington: *Francis Bacon*
Frank E. Manuel: *Isaac Newton, Historian*

The leading British authority on Montesquieu is Robert Shackleton: on other aspects of eighteenth-century social theory, see

Basil Willey: *The Eighteenth-Century Background*
Charles Frankel: *The Faith of Reason*
A. O. Lovejoy: *Essays in the History of Ideas*
Social Contract: ed. Sir Herbert Barker
C. L. Becker: *The Heavenly City of the 18th-Century Philosophers.*

6

Time's Creative Hand

FOR some time after the birth of the new sciences, a static vision of Nature and Society kept its hold on men's minds, even though the evidence needed to re-fashion it on new, historical foundations was beginning to accumulate. Yet this static vision persisted only in the same way that a super-saturated solution keeps its homogeneity and liquidity. The condition is essentially unstable: only a minor stimulus is needed—a few small crystals dropped into the solution—and the transformation is at once precipitated. One cannot pick out any single scientist or historian or philosopher to whom all subsequent historical and evolutionary attitudes can be traced back. By about 1800, historical attitudes and ideas were ready to crystallize out in many fields from the matrix of eighteenth-century thought, as men began for the first time to recognize the full extent of time and the crucial importance of development.

Vico: the Mendel of History

Lacking any one discoverer of historical development, accordingly, we must locate the origins of the new attitudes in another way: by putting our fingers on a few salient points within the eighteenth-century debate where the beginnings of a change can be felt, first in human history and then in the natural sciences. One finds the earliest of these significant starting-points around 1725 in the Kingdom of Naples—where Lorenzo Valla had done his own pioneer work on historical method nearly three centuries before. In this setting Giambattista Vico, a professor of jurisprudence and one of the most curious and isolated visionaries of our whole intellectual tradition, conceived for himself a novel approach to all aspects of human society. Vico referred to this method as his *Scienza Nuova*; and, though little was immediately done to follow up his intellectual programme, his books were the vehicle by which the concept of historical development at last entered the thought of Western Europe.

Contemporary ideas about history and society provided no more of a place for Vico's novel conceptions than Mendel's ideas were to find within biology more than a century later. Though Vico achieved a certain local fame at Naples itself, and the existence of his books was known more widely—his friend the Abbé Conti recommended them to Montesquieu—there is no conclusive evidence that he had any direct influence before the nineteenth century. It is not hard to see why. Even those features of his theory which appear to us most forward-looking must have struck his contemporaries as reactionary. For his fundamental ambition, which he did nothing to disguise, was to undo the intellectual damage done by Descartes, and to reinstate—though on a new and more profound basis—the older picture of human history as the continuous creative action of a divine Providence.

Vico knew very well that he could not pick and choose: he was not free to hang on to some parts of the Cartesian system, while rejecting others. Instead, he resolved to be rid of it entirely, and he began his demolition by attacking it at the keystone. Descartes, like Plato, had tried to rescue natural philosophy from scepticism by basing it on the timeless certitudes of geometry: he convinced himself that, if only the theories of natural science could be grafted on to mathematical axioms, some of this superior certainty might be transferred to them also. Vico now countered Descartes even more radically than Aristotle had done Plato. In our speculations about the world of Nature, he retorted, the kind of certainty which attaches to mathematical theorems is out of the question: consequently, Descartes' whole programme had been misconceived from the outset. Granted, the axioms and theorems of a well-formed mathematical system do present themselves to our minds with a unique clarity and certainty. But this is not—as Descartes had claimed—because God obligingly formed us with ideas which harmonize with the true realities of Nature. Mathematics is completely transparent to our minds, simply because it is *our arbitrary creation*. One can achieve a complete intellectual grasp only of things one has created oneself. We have made mathematics, but Nature was created by God. So perfect scientific certainty could be possessed by God alone and, in attempting to find a guarantee for his physical theories in the axioms of human mathematics, Descartes had been deceiving himself.

Vico did not return to the scepticism of Montaigne: the methods of observation, classification and hypothesis could give one a genuine, though modest, 'outside' knowledge of the material world, and mathematical techniques themselves might play a part in this process. But such 'outside' knowledge could never be more than partial and provisional. It could never be final, clear and certain, and Descartes' conviction that the results of mathematical physics were more certain than those of human history rested on a misconception. In fact, the contrary was the truth. The World of Nature is not a human creation, but the World of Nations is; so we approach an understanding of human societies as it were from the inside. In this way, history acquires a greater certainty and transparency than physics can ever have.

> In the night of thick darkness which envelopes the earliest antiquity, so remote from ourselves, there shines the eternal and never-fading light of a truth beyond all question: that the world of human society has certainly been made by men, and its principles are therefore to be found within the modifications of our own human mind.

The historian has to project himself back into the conditions of life current at any earlier stage, and see the problems of those days through the eyes of the men then living. He can then recognize which theories are congruous, or incongruous, with everything else we know about the period. This step made the principle of 'synchronisms' and 'anachronisms' the prime axiom of all historical enquiry.

In its content, Vico's account of the development of society is significantly like that put forward in antiquity by the Epicureans. With one qualification, the various 'ages' making up human history are not just variations on a theme: they are successive stages of a continuing process, each dependent on and developing out of the previous phase. The first men were semi-bestial giants, lacking the wits required to execute a 'social contract'. Social institutions came into being only slowly and painfully, laws being gradually imposed on men as the by-products of savagery and tyranny. But in spirit Vico's account was very different from the Epicureans': where Epicurus saw the effects of chance, Vico saw the hand of Providence, exploiting men's selfish preoccupations to promote wider purposes than they ever consciously envisaged. Yet, despite this difference of tone, the end-products are strikingly similar. On both accounts, the original anarchy was overcome only slowly and without human foresight, arts, crafts and social institutions being acquired one by one, as the inadvertent—and hence, for Vico, providential—results of human selfishness.

Vico's new historical method demanded an acute sensitivity to the nuances of language and the inner significance of myths. He stripped the picturesque surface off the older mythologies, to uncover the living thought that gave them birth. Originally, myths were neither poetic fancies, nor fictions of the priests, nor heroic legends magnified through the lens of the past. They represented rather Man's first crude but honest efforts to understand the world of Nature and live in harmony with it.

> In their fables the nations have, in a rough way and in the language of the senses, described the beginning of the world of the sciences, which the specialized studies of the scholars have since clarified for us by reasoning and generalization.

If mythological thought was anthropomorphic and animistic, that was due neither to perversity nor to affectation: it was only natural. Mythology was the thought of men living primitive lives, and was itself consequently primitive. For they could measure the world of Nature only by that which they already knew—namely, themselves.

> When men are ignorant of natural causes producing things, and cannot even explain them by analogy with similar things, they attribute their own nature to them. So the vulgar, for instance, say that the magnet loves iron.

In the 1720s, such insights were not merely original: they were beyond the grasp of most of Vico's contemporaries. Stacked together in the paragraphs of the *Scienza Nuova*, as in the dark and dusty passages of some over-stocked antique shop, we can find many of the fundamental axioms of twentieth-century anthropology, comparative law, literature and religion—and even linguistic philosophy.

Not only did Vico have a new sense of the organic links inter-connecting all aspects of a 'period', but he also freed himself entirely from the conventional eighteenth-century assumptions about the fixity of human nature and universal law. He argued that the principles of law and the social order were not permanent reflections of an unchanging human character, but the provisional products of a continuing historical process. All ages and societies re-create law in the context of their own cultures. Civilized society builds on the ruins of less civilized: its legal ideas would be unintelligible to the first, semi-bestial humans. Instead of seeing the ultimate source of political obligation in a mythical 'contract' too sophisticated for primitive men to understand, we should base our respect for the social order on a grateful understanding of the distance that Providence (or history) has raised us above that primitive anarchy.

Even the briefest account of Vico's ideas must include one qualification, which we must now add. Though in many respects his methods of thought looked forward to the future, the intellectual situation in which he exercised them was—inevitably—that of the 1720s. Like Hooke and Steno before him, Vico had somehow to reconcile his dynamic vision of the past with the short time-scale of Biblical chronology. (His own calculations assumed a Creation in 3995 B.C.) In fact, he had to compromise, and he did so quite as cheerfully as the geologists. Our historical enquiries, he declared, were subject to two limitations. First, apart from what the Old Testament told us we could know nothing at all about the conditions of life before the Flood, which he dated to 2339 B.C., since it had obliterated all traces of the antediluvian civilization. Furthermore, the fragmentation of the human race following the incidents at the Tower of

Babel affected different peoples in different ways. While all the Gentile races were afflicted, the Jews were spared. God's will having been revealed to the Jewish people, they preserved a moral understanding and a political order when all other men were scattered abroad as undisciplined, ignorant savages. So Vico's theory of historical development did not apply to the Jews: it was a theory of 'Gentile history'. Yet, though the Jewish scriptures were sacred and the commandments of Moses divinely-ordained, this very fact made them—paradoxically—less interesting rather than more so, for it removed them beyond the reach of human understanding. It was Gentile history that interested Vico—that gradual, painful ascent through a sequence of stages, each with its own laws, customs, styles of language and modes of thought.

Such a man, with such a vision, could scarcely commend himself to the intellectuals of eighteenth-century France. They saw social progress as dependent on men's discovering the fixed laws of human nature, and consciously exploiting this knowledge, so substituting reason and enlightenment for emotion and superstition. In Vico's eyes, there were no such fixed laws to be understood, nor did civilized society result from conscious human effort. It was a providential and uncovenanted by-product of the interplay between conflicting self-interests. So Vico and his contemporaries hardly even shared a common vocabulary. Yet, buried beneath the vast glacier of eighteenth-century thought, Vico's message remained dormant but alive: once the intellectual thaw began, the vision of 'the living force of humanity creating itself' (as his advocate Michelet summarized it) re-emerged to take a central place in nineteenth-century historical analysis and social theory.

Kant and Cosmic Evolution

Vico's attack on a static conception of human nature did not extend to the physical world; for the very foundation of this attack lay in the conviction that human activities have their own 'inner' motivation, which we can reconstruct by an effort of sympathetic imagination. In his eyes, historical development was a peculiarly human phenomenon, to which there was no counterpart in the world of Nature. The idea of interpreting the whole universe as the product of a historical development originated quite elsewhere: in astronomy rather than social theory, and on the Baltic rather than the Mediterranean. Curiously enough, both the major revolutions underlying the modern conception of Nature—the expansion of Space, then of Time—had roots in East Prussia, that frontier region around Königsberg (or Kaliningrad) so long in dispute between Germans, Poles and Russians. Copernicus himself was a Canon of Frauenburg

(Frombork), while Kant and Herder both grew up and worked at Königsberg.

The fame of Immanuel Kant's three *Critiques* has obscured his striking contributions to cosmology. In fact, his earlier work on the *General History of Nature and Theory of the Heavens* (1755) was the first systematic attempt to give an evolutionary account of cosmic history: in it, he spoke of the whole Order of Nature, not as something completed at the time of the original Creation, but as something *still coming into existence*. The transition from Chaos to Order had not taken place all at once. On the contrary:

> The future succession of time, by which eternity is unexhausted, will entirely animate the whole range of Space to which God is present, and will gradually put it into that regular order which is conformable to the excellence of His plan. . . . The Creation is never finished or complete. It did indeed once have a beginning, but it will never cease.

Kant justified this original and heterodox conclusion by a sweeping extension of the Newtonian idea of gravitation. Newton himself had explained free fall, the tides, and the movements of the comets and planets in gravitational terms. Kant now applied the theory more speculatively and more extensively, to embrace the structure of the Milky Way, and even the movements of other galaxies far removed from our own. Gravity became the great instrument by which God was slowly imposing order on the unbounded material universe.

Kant acknowledged his debt to Thomas Wright, whose account of the Milky Way had been published five years earlier:

> The systematic arrangement which was found in the combination of the planets which move around the Sun, seemed in the view of astronomers . . . to disappear in the multitude of the Fixed Stars; and it appeared as if the regulated relation which is found in the smaller Solar System, did not rule among the members of the Universe as a whole. The Fixed Stars exhibited no law by which their positions were bounded in relation to each other; and they were looked upon as filling all the Heavens and the Heaven of Heavens without order and without design. . . .
>
> It was reserved for an Englishman, Mr. Wright of Durham, to make [the next] happy step. . . . He did not regard the Fixed Stars as a mere swarm scattered without order and without design, but found a systematic constitution in the whole Universe and a general relation of these stars to the ground-plan of the regions of Space which they occupy.

The Milky Way, Kant explained, was an 'indefinitely enlarged' counterpart of the Solar System. But this was only a beginning. If one now supposed that our own galaxy is not unique, the scope of the gravitational theory could be extended much farther:

> If a system of Fixed Stars which are related in their positions to a common plane, as we have delineated the Milky Way to be, be so far removed from us that the individual stars of which it consists are no longer sensibly distinguishable even by the telescope; [and] if such a World of Fixed Stars is beheld at such an immense distance from the eye of the spectator situated outside of it, then this World will have the appearance of a small patch of space, whose figure will be circular if its plane is presented directly to the eye, and elliptical if it is seen from the side or obliquely. The feebleness of its light, its shape, and the apparent size of its diameter will clearly distinguish such a phenomenon, when it presents itself, from all the stars that are seen single.

'This phenomenon', he goes on, 'has been distinctly perceived by different observers.' In 1742, for example, Maupertuis had described certain 'nebulous' stars: 'small luminous patches, only a little more brilliant than the dark background of the heavens; visible in all parts of the sky; presenting the figure of more or less open ellipses; their light being much feebler than that of any other object we can perceive in the heavens'. Kant at once identified these 'luminous patches'—quite correctly, on our modern view—as being other galaxies, strictly comparable to our own.

> As these nebulous stars must undoubtedly be at least as far away from us as the other Fixed Stars, it is not only their magnitude which would be so astonishing—seeing that it would necessarily exceed that of the largest stars many thousand times: strangest of all would be the fact that self-luminous bodies or Suns of this extraordinary size show still only the dullest and feeblest light.
>
> It is far more natural and intelligible to regard them as being not enormous single stars, but systems of many stars, whose distance presents them in such a narrow space that their light, which is individually imperceptible, reaches us, on account of their immense number, as a uniform pale glimmer. . . . And this is in perfect harmony with the view that these elliptical figures are just Universes or (so to speak) Milky Ways, like those whose constitution we have just unfolded.

From now on, men had to reckon with an unlimited number of galaxies, scattered throughout Space, and even more remote from one another than the stars within our own galaxy.

Kant did not find this new perspective either daunting or unintelligible. He believed that it could readily be explained on Newtonian principles, and furnished 'an idea of the work of God which is in accordance with the Infinity of the great Architect of the Universe'. Gravity maintained the same coherence between the different stars within each galaxy that it did with the Solar System; and, moreover, the same 'systematic connection' presumably bound all the galaxies together. So, by 'the connecting power of [gravitational] attraction and centrifugal force', the totality of nature became a single physical system.

When Kant applied this theory to the origin of the cosmos, he transformed the traditional idea of Creation entirely. In the beginning, the raw material of the universe was distributed randomly, without form or order, but subject to the regular Newtonian forces. Mathematically speaking, the cosmos was infinite, so it had no geometrical 'centre'; but, since the primaeval matter was spread randomly rather than uniformly, the point at which this raw material *happened* to be most densely accumulated became the physical focus or 'centre' around which the Order of Nature began to establish itself.

> At the primary stirring of Nature, formation will have begun nearest [to this Centre of densest accumulation]; and in advancing succession of time the more distant regions of Space will have gradually formed worlds and systems with a systematic constitution related to that Centre. Every finite period, whose duration has a proportion to the greatness of the work to be accomplished, will always bring only a finite sphere to its development from this Centre; while the remaining infinite part will still be in conflict with the confusion and chaos, and will be further from the state of completed formation, the further its distance is from the sphere of the already developed part of Nature.

This gradual establishment of order out of chaos had taken a vast period of time:

> There had perhaps flown past a series of millions of years and centuries, before the sphere of ordered Nature, in which we find ourselves, attained to the perfection which is now embodied in it; and perhaps as long a period will pass before Nature will take another step as far in chaos. But the sphere of developed Nature is incessantly engaged in extending itself. Creation is not the work of a moment. . . . Millions and whole myriads of millions of centuries will flow on, during which always new Worlds and systems of Worlds will be

> formed, one after another, in the distant regions away from Centre of Nature, and will attain to perfection. . . .
>
> This infinity and the future succession of time, by which Eternity is unexhausted, will entirely animate the whole range of Space to which God is present, and will gradually put it into that regular order which is conformable to the excellence of His plan. . . . The creation is never finished or complete. It has indeed once begun, but it will never cease. It is always busy producing new scenes of nature, new objects, and new Worlds. The work which it brings about has a relationship to the time which it extends upon it. It needs nothing less than an Eternity to animate the whole boundless range of the infinite extension of Space with Worlds, without number and without end.

The eternal process of cosmic creation was thus identified with the evolution of Nature.

This was the voice of a new age. The notion of continuous creation—a cosmic evolution inexhaustible in any finite period of time—was as visionary and imaginative as anything in the writings of Vico. Kant now deliberately set on one side Newton's careful distinction between the creation of the present Order of Nature and its maintenance: the only creation we need demand was the progressive victory of order over chaos, throughout an infinity of time. Though Kant's religious piety is unquestionable, yet his cosmology flouted the chronology of the Bible openly and completely. By 1750, men could contemplate a future lasting many thousands of years; but no one before Kant had talked so publicly and seriously of a *past* comprising 'millions of years and centuries'. If his book escaped ecclesiastical condemnation, this was owing less to the orthodoxy of its contents than to the protection of Frederick the Great and the bankruptcy of his publisher. For the time being, its influence was local and limited, and after Frederick's death Kant was forced to recant. As with Vico's *New Science*, though the existence of the book was known, the force of its argument was appreciated only in the nineteenth century.

Kant's cosmology clearly owed more to Fontenelle's *Plurality of Worlds* than it did to the Old Testament. In the balance between Biblical interpretation and physical science, the scale was at last beginning to tilt decisively in favour of physics. Whereas Burnet had still taken the framework of his cosmic history from the Bible story, Kant took his fundamental data from dynamics and astronomy, and used them to draw as edifying and orthodox a picture as he could contrive. Fontenelle's Marquise had expressed consternation at the prospect of suns and stars being extinguished, but Kant found in their very mortality fresh evidence of the Creator's hand, working through the fecundity of Nature:

> The inevitable tendency which every World that has been brought to completion gradually shows towards its destruction, may even be reckoned among the reasons which may establish the fact that the Universe will again be fruitful of Worlds in other regions, to compensate for the loss which it has suffered in any one place. The whole portion of Nature which we know, although it is only an atom in comparison with what remains concealed above or below our horizon, establishes at least this fruitfulness of Nature, which is unlimited, because it is nothing else than the exercise of the Divine omnipotence.
>
> Innumerable animals and plants are daily destroyed and disappear as the victims of time; but nonetheless Nature, by her unexhausted power of reproduction, brings forth others in other places to fill up the void. Considerable portions of the Earth which we inhabit are being buried again in the sea, from which a favourable period had drawn them forth; but at other places Nature repairs the loss, and brings forth other regions which were hidden in the depths of being in order to spread over them the new wealth of her fertility. In the same way, Worlds and systems perish and are swallowed up in the abyss of Eternity; but at the same time Creation is always busy constructing new formations in the Heavens, and advantageously making up for the loss.

In a grandiose peroration, Kant looked forward to a Stoic *Götterdämmerung*, in which all the matter of our galaxy would fall together in a violent conflagration, only to be dispersed once more into the surrounding space by 'the ensuing unspeakable heat' and so 'furnish materials for new productions by the same mechanical laws, whereby the waste space will again be animated with worlds and systems'.

> When we follow this Phoenix of Nature, which burns itself only in order to revive again in restored youth from its ashes, through all infinity of times and spaces; when it is seen how Nature, even in the regions where it decays and grows old, advances unexhausted through new scenes, and, at the other boundary of Creation in the space of the unformed crude matter, moves on with steady step, carrying on the plan of the Divine revelation, in order to fill Eternity, as well as all the regions of Space, with her wonders: then the spirit which meditates upon all this sinks into profound astonishment.

Kant's message of pious wonder in the face of Eternity is summed up in some verses which he quotes from the poet-physiologist, Albrecht von Haller:

Infinity! what measures thee?
Before thee worlds as days, and men as moments flee!
Mayhap the thousandth sun is rounding now;
And thousands still remain behind!

Even as the clock its weight doth wind,
A Sun by God's own power is driven;
And when its work is done, again in heaven
Another shines. But thou remainst! to thee all numbers bow....

And when the World shall sink, and Nothing be once more,
When but its place remains, and all else is consumed;
And many another heaven by other stars illumed,
Shall vanish when its course is o'er:
Yet thou shalt be as far as ever from thy death,
And as today thou then shalt breathe eternal breath.

Herder and the Development of Nature

Vico and Kant had both taken crucial steps towards a more dynamic view, recognizing that the world was the transient product of a continuous process of development still capable of producing genuine novelties. But each man had worked only at one end of the scale: Kant applying this new vision to the astronomical macrocosm, Vico to the microcosm of human life and society. Their ideas were amalgamated by Johann Gottfried Herder, who learned his natural science at Königsberg from Kant himself, but derived his political and social ideas more from Kant's colleague Hamann—one of the few political philosophers before 1800 to have any sympathy for Vico.

Reconciling Vico and Kant was not entirely straightforward, for in certain respects their positions were sharply opposed. Vico believed that genuine historical novelties could arise during the evolution of society, just because it was a human creation: there was no corresponding progress in Nature. For Kant, it was the other way round. The purely material aspect of Nature was continuously evolving, as systematic order progressively extended into unbounded Space; but this was the full extent of his evolutionary ideas. Any hint of genealogical relationships between different species, for example, filled Kant with distaste: in his view, they had existed as distinct and mutually infertile stocks ever since their creation.

A *relationship* between them—such that one species should originate from another and all from one original species, or that all should

> spring from the teeming womb of a universal Mother—this would lead to ideas so monstrous that the reason shrinks from before them with a shudder.

Human law and morals, too, were—or should be—governed by unchanging ideals of rationality. True: there was a certain pattern of progress in human history since its scriptural origins. Men had not always conformed to the permanent ideals of reason, and their irrational motives had been exploited by Nature, with the result that the original life of anarchy had been displaced by a more rational, law-governed order of society:

> Man desired concord; but nature knows better what is good for his species; she desires discord. Man wants to live easy and content; but nature compels him to leave ease and inactive contentment behind, and throw himself into toils and labours, in order that these may draw him to use his wits in the discovery of means to rise above them.

Still, the ideals implicit in this historical process were themselves *fixed* ones, arising out of Man's innate rational capacity, and the 'hidden plans of nature' exemplified in history were as definite and discoverable as the 'laws of nature' studied in physical science.

Herder rejected Kant's social ideals as too static. Man's potentialities would never be realized consciously, deliberately or finally: the course of human history was just as much a continuing, creative and eternally unfinished process as Kant's cosmic evolution. From 1784 on, he presented his own account of cosmic development, in the four volumes of his *Ideas towards a Philosophy of the History of Man*. But this time the historical process embraced the stars and planets, the Earth, living species and human society alike. Herder's *Ideas* constituted a philosophical vision rather than a scientific theory, and his attitude can easily be criticized as romantic, sweeping and high-flown. Yet his system had a widespread influence. In its own way, the older, static vision of Nature and Society had been equally sweeping, and there was some virtue, even for scientific purposes, in showing that an alternative, progressive approach could be adopted consistently, in both the natural and the human sciences.

Like most transitional theories, Herder's is a compromise between the old and the new. He presents a picture of the cosmos as the product of historical development, but he describes this development in providential, even theological, terms. Beginning with the development of the galaxies, he focusses in succession on the planetary system, the Earth, living things, Man, and finally the European society of his own time. This order of exposition has both an intellectual and a temporal significance: as the attention is progressively restricted, the aspects of the cosmos considered become more localized and more recent in origin. First the galaxy came into existence, then the Sun and the system of planets: of these, the Earth provided the fittest environment for the appearance of living things, being neither too hot nor too cold. On the Earth, all the creatures in Aristotle's Ladder of Nature had appeared in succession:

> From stones to crystals, from crystals to metals, from these to plants, from plants to animals and from animals to man, we see the form of organization ascend; and with it the powers and propensities of the creature become more variant, until finally they all, so far as possible, unite in the form of man.

In turn, each species provided an essential part of the environment for those which were created later: the Great Chain of Being, which up till then had been a timeless pattern of reality, acquires a time-dimension—the intellectual passage from 'lower' to 'higher' now reflects a succession from 'earlier' to 'later'. And the whole ordered cosmos, embracing Nature and Society alike, is unified by a harmonious system of 'forces':

> The power which thinks and operates in me is in its nature a power as eternal as that which holds together suns and stars: wherever and whoever I shall be, I shall be what I am now, a force in the system of all forces, a being in the immeasurable harmony of the world of God.

Some people have seen in the time-dimension of Herder's zoology evidence that he anticipated Darwin. But this is a mistake, for in his argument there is no suggestion that later and higher animals were descended from earlier and lower ones. He merely points out that the existence of simpler organisms was an essential condition for the survival of more complex forms.

> It is manifestly contrary to Nature that she should bring all creatures into existence at the same time. The structure of the earth and the inner constitution of the creatures themselves make this impossible. Elephants and worms, lions and infusoria, do not appear in equal numbers, nor could they be created, in consistency with their natures, at one time or in equal proportion. Millions of shellfish must needs have perished before our bare rock of earth could be made a fruitful soil for a finer type of life; a world of plants is destroyed each year in order that higher beings may be nourished thereby. Even if one wholly disregards the final causes of the creation, yet even in the very raw material of Nature there lies the necessity that one being

> should come out of many, that in the revolving cycle of creation countless multitudes should be destroyed so that through this destruction a nobler but less numerous race might come into being . . .
>
> Man, therefore, if he was to possess the earth and be the lord of creation, must find his kingdom and his dwelling-place made ready; necessarily, therefore, he must have appeared later and in smaller numbers than those over whom he was to rule.

This distinction is important. Organic evolution, as Darwin conceived it, involved at least three distinct propositions: first, that more complex forms of life appeared on the Earth later than simpler ones (the doctrine of *progression*); secondly, that these later forms of life were descended from the earlier ones (the doctrine of *transformation*); and thirdly—Darwin's essential contribution—that the descent of these later species from the earlier was a consequence of variation and natural selection. Like many of Darwin's immediate predecessors, Herder was a 'progressionist' but not a 'transformist', and he explicitly rejected the idea of evolution by descent—contradicting, in particular, Monboddo's assertion that apes and men belonged to a common species. One could, of course, also be a transformist without being a progressionist, like Robert Hooke; or a believer in natural selection without transformation, like Empedocles and Lucretius. Between 1650 and 1850, men argued for most of the possible combinations of views, and they were finally reconciled to the Darwinian position only after all the others had been laboriously eliminated.

Once the human race was established on Earth, Herder argued, the 'purposes of Mother Nature' could be carried further. Through his upright posture, Man became Lord over the Animal Creation (*der Gott der Tiere*) and embarked on the next phase of cosmic development—realizing his own potentialities through the creation of society. In the different climates and continents, the various branches of the human race grew up with different characters and capacities: in succession, the historical cultures realized the special potentialities of each nation and society. Whereas, for Herder's predecessors, human history had illustrated eternal truths about human nature, it now became a study of the *progressive development* of that nature, in which later achievements took advantage of, and built on earlier ones:

> That this march of time has had an influence on the mode of thought of our race is undeniable. Seek now, or attempt, an Iliad, try to write as Aeschylus, Sophocles and Plato did: it is impossible. The simple, childlike mind, the untroubled outlook on the world—in short, the Hellenic *Jugendzeit*—is a thing of the past. Likewise with the

> Hebrews and the Romans: on the other hand, we know and understand a multitude of things of which both Hebrews and Romans were ignorant. The one had had a day's teaching, the other a century's. Tradition has been enriched, the Muse of the Age—History itself—speaks with a hundred voices and sings with a hundred tones. Let it be granted that, in the immense snowball which the movement of time has rolled up for us, there may be included as much folly and confusion as you please: nevertheless, even this confusion is the offspring of the centuries, and could have arisen only from the indefatigable advance of that same being. Any return to former times (even Plato's famous Great Year) is, on this conception of the cosmos and of time, an impossibility. We are carried ever forward; the stream never returns to its source.

More than a century later, the National Socialists of the German Third Reich quoted Herder to justify the exploitation of inferior 'non-Aryan' peoples by their own 'master-race'. But this was a cruel distortion of his own more tolerant internationalism—

> The historian of mankind must, like the Creator of our race or like the genius of the earth, view without partiality and judge without passion. . . . Nature has given the whole earth to her human children and has permitted all to germinate upon it that, by virtue of its place and time and potency, could germinate. All that can be, is; all that can come to be, will be; if not today, then tomorrow . . .
>
> Nature's year is long; the blooms of her plants are as many as these growths themselves and as the elements that nourish them. In India, Egypt, China, that has come to pass which nowhere and never will again come to pass on the earth; and so in Canaan, Greece, Rome, Carthage. The law of necessity and congruity, which is composed of potencies and place and time, everywhere brings forth different fruits.

So, in the ideas of Herder, the Great Chain of Being merged into a Chain of Cultures, in which social development was justified as a continuation of cosmic history by other means. At this point, we may for the moment leave the general philosophy of history. Herder's attempt to unite Nature and Society in a single process of development was not the last word on the subject. Some of his successors rejected this dovetailing of such different processes: Hegel, for instance, preferred to revive Vico's contrast between the repetitive processes of Nature and the progressive march of human history; Marx and Engels, in their turn, reasserted the continuity between Nature and Society; and on that level the debate is still in progress. By the end of the eighteenth century, however, the *philosophical* ground had been cleared of absolute prejudices against the idea of development, and men were taking a new and more empirical look at the History of Nature itself. Though questions about the origin of the Earth could still run foul of the Old Testament narrative, the Biblical chronology was dying on its feet. The more empirical discoveries men made in the fields of astronomy, physics, chemistry and natural history, the more coherent and consistent a picture of the past they could build up; and a point soon came at which this picture broke out of the six-thousand-year time-scale of Eusebius and Julius Africanus.

FURTHER READING AND REFERENCES

Admirable modern editions of G. B. Vico's *Autobiography* and *New Science* have been prepared by Bergin and Fisch: for other commentaries on Vico, from an idealist standpoint, see

B. Croce: *The Philosophy of Giambattista Vico*
R. G. Collingwood: *The Idea of History*

For a brief summary of Vicos views, it is well worth hunting out Isaiah Berlin's masterly essay on 'The Philosophical Ideas of Giambattista Vico', in the collection *Art and Ideas in Eighteenth-Century Italy*, published under the auspices of the *Istituto Italiano di Cultura*

For Kant's cosmological ideas, see

W. Hastie: *Kant's Cosmogony*
Theories of the Universe: ed. M. Munitz
John C. Greene: *The Death of Adam*

There are illuminating discussions of Herder's ideas in

A. O. Lovejoy: *Essays in the History of Ideas*

The biological views of Kant and Herder are discussed in excellent essays by Lovejoy in the collection

Forerunners of Darwin: ed. Glass, etc.

All three philosophers are represented in the collection

Theories of History: ed. Patrick Gardiner.

7

The Earth Acquires a History

DURING the second half of the eighteenth century, philosophical debate had shown the possibility of replacing the static Renaissance view of Nature by a dynamic, developmental one. But philosophy alone could not dispatch the older time-scale finally: for that, an alternative numerical chronology had to be worked out from the evidence of scientific observation. Now it was up to the natural scientists to carry this possibility into effect, and to demonstrate how the natural world could have acquired its present form through the lapse of time. So, by A.D. 1800, geologists and zoologists were once again attacking the fundamental questions posed originally by Anaximander and his fellow-Ionians, at Miletos some 2300 years before. In the event, the first branch of natural science to become genuinely historical was geology: the crucial battle between scriptural chronology, based on human traditions, and natural chronology based on 'the testimony of things', was fought out over the history of the Earth. If our ideas about the past are now no longer restricted within the time-barrier of earlier ages, this is due above all to the patience, industry and originality of those men who, between 1750 and 1850, created a new and vastly extended time-scale, anchored in the rock strata and fossils of the Earth's crust.

This is not to say that the *motives* of the first modern geologists were universally, or even predominantly, historical ones. To begin with, many of them were concerned only to study the present make-up of the Earth's crust, classify the different rocks, and see whether any consistent order could be found in the strata of different countries and regions; and often enough the reasons for their interest in such questions were practical ones —for geology has always been largely an applied science, allied to mining and metallurgy. The historical significance of their new knowledge dawned on them only gradually. Initially, they scarcely recognized that the Earth had a history at all: at any rate, anything more than was contained within the story of the Creation and the Deluge. But, even before 1800, discoveries were accumulating which, on the older view, were frank anomalies. Some of these, like the fossil marine shells found in inland rocks, had been known since the classical era: others, like the extinct volcanoes of the French *Massif Central*, were now recognized for the first time. Either way, these anomalies served as pin-pricks to men's confidence in the Creation-story and the accepted time-scale. Unless those were called in question, it was equally embarrassing to explain the geological origin of the anomalous formations, or to dismiss them as incomprehensible survivals from the original Creation. Once the fact of geological change had been admitted, questions about the temporal sequence of these changes were inescapable: what agencies were responsible, whether they were the same as those now acting, how long they had taken to produce their visible effects. In their turn, these historical questions led to further research, and so to more discoveries, which rebounded once again on inherited ideas and assumptions. So observation and theory snowballed, and a brand-new temporal framework was gradually forced on men.

During the hundred years of this progressive erosion and resynthesis, men lived with uncertainties as profound, and dilemmas as agonizing, as those provoked earlier by Copernicus' reforms in astronomy. In some ways, the situation was even more difficult. The new geology could never hope to achieve the kind of mathematical certainty which Newton's theories stamped upon the new astronomy: like a successful piece of criminal detection, it carried conviction only through the cumulative weight of circumstantial evidence. And, as in the Copernican revolution, though the outlines of the new system were discerned early, it superseded the older one only when it had been worked out in detail and established unanswerably. In such an intellectual situation, this type of pattern is probably inevitable. Where the issues are so profound, it is not enough to throw doubt on older ideas. An alternative account must be presented which is equally complete and more convincing; and, until this is done, men cannot be blamed for being tenacious of long-cherished traditions.

We cannot hope in this volume to cover adequately the whole development of geology between 1750 and 1850. (The subject has in any case been admirably handled in a number of recent books, referred to below.) Instead, we shall concentrate on those aspects of the story which are directly relevant to our larger theme: how men acquired a sense of history, and realized the extent of past time.

The Epochs of Nature

By 1750, men were beginning to recognize that the present face of the world might carry enduring traces dating from much earlier, even prehuman times; and that, if only we could interpret them, these traces would provide evidence about the past as direct and reliable as any human tradition. But the problem of interpretation was still as much of a challenge as ever. Human records might be corrupt, and so untrustworthy, but at any rate they were 'testimony' in the literal sense of the term: that is, undisguised assertions about the past. Rock-strata or fossils provided 'testimony' only in a transferred sense: their present forms served as clues to events in former epochs only for men whose scientific principles gave them a basis for interpretation. The question was still, as it had always been, 'What kind of principles can justify retrospective inferences, and so provide an intellectual bridge back to the past?'

The men who now attacked this question relied on two main lines of argument, one broader, the other narrower. Some attempted to fit geology into Newton's all-embracing framework of physical laws. Given the properties and dimensions of the planet on which we live, what (they asked) do the principles of Newtonian physics imply about its earlier history? Others proceeded in a more piecemeal way. By studying in detail the structure of the Earth's surface, they hoped to reconstruct the successive stages by which the Earth acquired its present form. If each different type of rock or fossil could be persuaded to tell its own story, these separate testimonies should combine to give a consistent account of the Earth's development. Later, a third line of argument was added, based on biological principles rather than physical ones, and the intellectual claims of the modern, extended time-scale were finally established by the resultant interweaving of geological considerations with evolutionary ones.

The first excursion into the physical history of the Earth to result in serious numerical estimates of past time was published in the 1770s by that striking, prolific and courageous figure, Georges Louis Leclerc, Comte de Buffon. Buffon had no hesitations. He never doubted that the Earth had in fact a long history, nor that the task of reconstructing it was within the capacities of the human reason, and his overall programme of enquiries has outlived all his detailed arguments. For he analysed with great elegance the scientific problems which the 'time-barrier' posed for natural historians, and he established convincingly that, even on the most restrictive assumptions, a geological history based on the established principles of physics would extend the scale of our chronology to 168,000 years, at the very least. (His own private estimate was nearer half a million years.)

Like Aristotle and Linnaeus, Buffon was less an individual scientist than a committee, and his library was the centre for a wide circle of correspondents. His *Natural History* was a life-work: the first three volumes were published in 1749, and the series was not complete when he died in 1788, just before the French Revolution. Buffon had aimed to produce a scientific encyclopaedia dealing in succession with the planetary system, the Earth, the human race and the different kingdoms of living creatures: at the time of his death, his published volumes had got as far as the birds, and he was in the middle of cataloguing the fishes. Described so baldly, his achievement may sound pedestrian, but in fact he saw this multitudinous mass of detail in relation to a wider—and an original—scheme of ideas. Furthermore, his books contain long and penetrating digressions about scientific theory and method, and so read at times like a scientific *Tristram Shandy*: any curious and intriguing fact of nature (say, the sterility of mules) is liable to be an occasion for a far-ranging theoretical argument. Still, behind these apparent irrelevancies, there is an underlying system. His books cover some forty years of a busy career, and these theoretical asides show his point of view developing from decade to decade, in the light of new discoveries and controversies.

Buffon's conception of *Histoire Naturelle* embraced what we are here calling the history of Nature. His early hypothesis about the origin of the Solar System was intended to be the opening instalment of a story which would cover cosmological, geological and zoological development. These plans were, however, frustrated by the theologians of the Sorbonne, and he was compelled to issue a formal retraction. For the next twenty-five years Buffon kept his unorthodox speculations to himself, but by 1774 he felt sufficiently secure to return to the dangerous topics. In his *Introduction to the History of Minerals*, he described a comprehensive series of experiments on the rates at which spheres of different sizes and substances cooled down; and in a long appendix he calculated the times needed for the different planets and their satellites to cool from a white heat to an inhabitable temperature. He wrote without apology, just as though his *Theory of the Earth* had never been condemned. His purpose was concealed partly by the laboriousness of his arithmetic, partly by a mollifying heading (*partie hypothétique*). Reassured by the reception of this *Introduction* he followed it in 1778 with a fuller account of the successive *Epochs of Nature* through which the Earth had presumably developed.

The fundamental problem for the history of Nature has never been more clearly formulated than it was in his opening words:

> Just as in civil history we consult warrants, study medallions, and decipher ancient inscriptions, in order to determine the epochs of the human revolutions and fix the dates of moral events, so in natural history one must dig through the archives of the world, extract ancient relics from the bowels of the earth, gather together their fragments, and assemble again in a single body of proofs all those indications of the physical changes which can carry us back to the different Ages of Nature. This is the only way of fixing certain points in the immensity of space, and of placing a number of milestones on the eternal path of time.

1. The Palermo Stone (*c.* 2500 B.C.): the oldest surviving written chronicle, recording the history of the Egyptian Kingdom year by year (see p. 26)

2 & 3. Early illustrations of geological formations (*above*) and fossil impression of a fern (*below*)
[from *Nature Display'd* (3rd edn., 1740)]

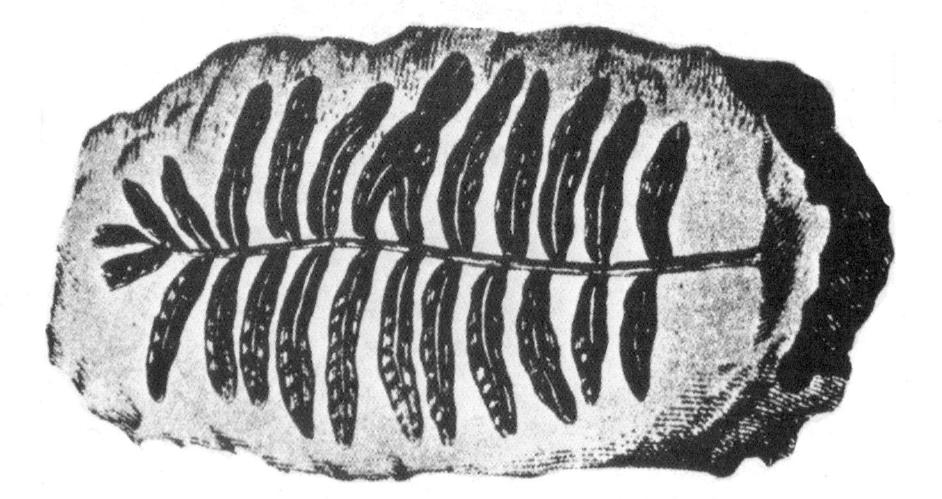

4. The Valley of Montpézat (Ardèche), showing a typical volcanic basalt stratum cut through subsequently by streams

[from G. P. Scrope, *The Geology and Extinct Volcanoes of Central France* (2nd edn., 1858)]

5. Fossil brachiopod (*Cyrtospirifer*) in Upper Devonian slate: age approx. 440 million years

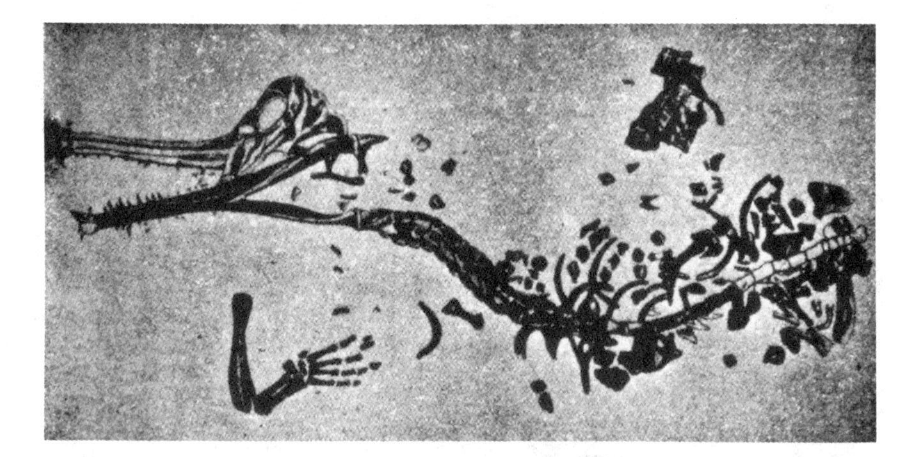

6. Engravings from Cuvier's *La Règne Animale*, showing (*above*) fossil remains of *Crocodilus Priscus* and (*below*) reconstruction of *Megatherium* (see p. 181)

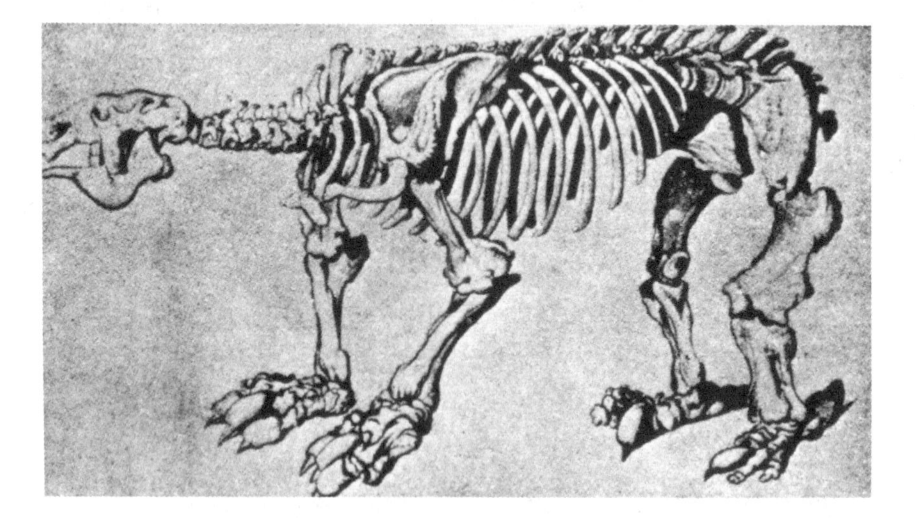

7. Darwin's finches, showing variations between the species found on the different islands of the Galápagos group (see p. 200)

[by courtesy of David Lack and the Cambridge University Press]

8. Cartoon from *Punch*, 25th May 1861 (see p. 120)

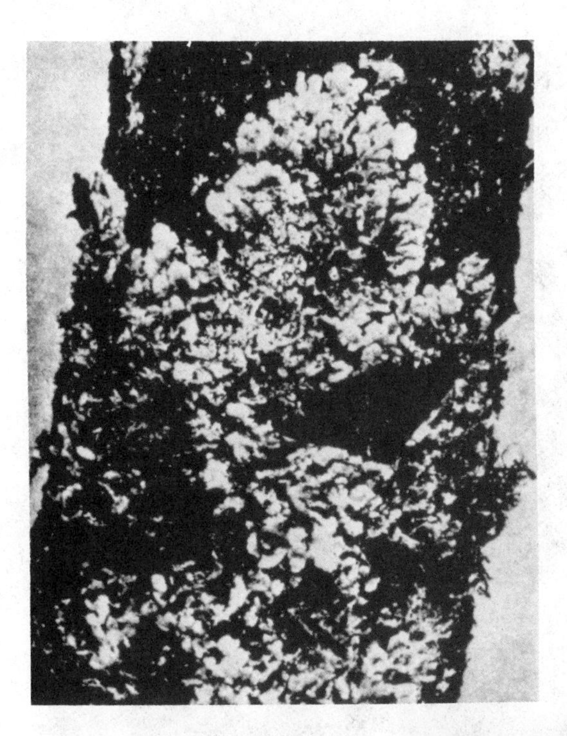

9. Light and dark form (*carbonaria*) of the Peppered Moth (*Biston betularia*) on lichen-covered tree-trunk in countryside (*above*), and on soot-covered tree-trunk near Birmingham (*below*)—see p. 220

[by courtesy of Dr. H. B. D. Kettlewell]

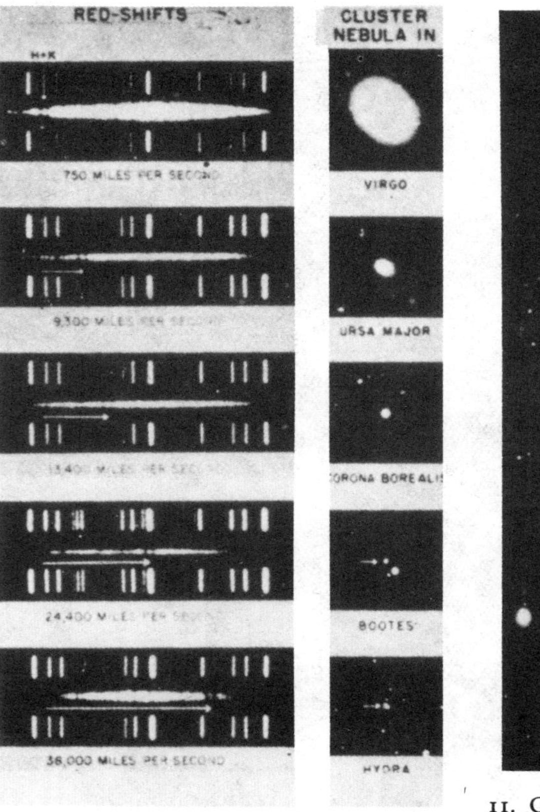

10. Relation of red-shift and distance for extra-galactic nebulae, showing shift in calcium absorption lines towards red end of spectrum. The estimated distances of the respective nebulae are in the proportion 1 (Virgo) : 13 : 17 : 31 : 47 (Hydra) : on the Döppler interpretation, these red-shifts correspond to the velocities with which the nebulae are receding—as indicated (see p. 255)

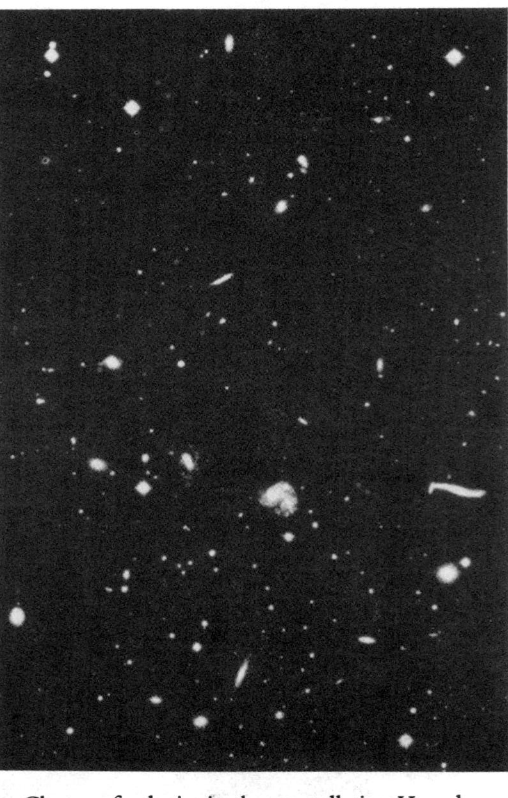

11. Cluster of galaxies in the constellation Hercules, photographed with the 200″ telescope (cf. p. 257)

> The past is like distance: our view of it would shrink and even be lost entirely, if history and chronology had not marked the darkest points by beacons and torches. Yet despite these lights of written tradition, let us go back a few centuries, and how uncertain are our facts! How confused are the causes of events! And what profound darkness enshrouds the periods before that tradition! Besides, it tells us only about the deeds of a few nations, i.e. the doings of a very small part of mankind: the rest of the human race is as nothing, either for us or for posterity. Thus civil history, bounded on one side by the darkness of a period not far distant from our own, embraces in the other direction only those small areas of the Earth which have been occupied in succession by peoples mindful of their own tradition. Natural history, on the other hand, embraces in its scope all regions of space and all periods equally, and has no limits other than those of the universe.

However unchanging Nature might appear to human eyes, the Earth's present state was without question very different from its original one; and, during the intervening period, it must have passed through several other phases, occupying longer or shorter times. These periods he called *epochs*, and the problem was to reconstruct them, using evidence of three different kinds:

> The surface of the Earth has taken different forms in succession; even the heavens have changed, and all the objects in the physical world are, like those of the moral world, caught up in a continual process of successive variations . . .
>
> But so as to pierce the night of past time—to recognize by a study of existing objects the former existence of those which have been destroyed, and work our way back to this historic truth of buried facts by the force of existing facts alone; so as to judge (in short) not merely the recent past, but also the more remote, on the basis of the present alone . . . we shall employ three prime resources: (i) those facts which can take us back to the origin of Nature; (ii) those relics which must be regarded as evidence of earlier eras; (iii) those traditions which can give us some idea about subsequent eras—after which we shall attempt to link them all together by comparisons, so as to form a single chain descending from the zenith of the scale of time down to ourselves.

By 'facts', he meant those physical properties on which he founded his arguments about the cooling of the planets; by 'relics', he meant such things as fossils, shells and mammoth-bones; and, by referring also to 'traditions', he hoped to forestall the inevitable religious opposition—attempting to harmonize his scientific ideas with the Old Testament, by judiciously reinterpreting the opening chapters of *Genesis*. The *Book of Genesis* (he argued) had not been written for scientists, but for the unlearned. The Days of Creation could not have been 'days' such as we know now—of twenty-four hours each—since the very succession of day and night was established only on the third 'day', after the creation of the Sun. Rather, we should read the Biblical word 'days' as referring to periods of indefinite length, about whose exact duration Moses had not committed himself. In this way, we could rescue *Genesis* from all danger of contradicting the facts of Nature and the conclusions of reason.

With this preamble, Buffon launched into an account of the geological epochs corresponding to the seven Days of Creation. These epochs had lengths ranging from 3000 to 35,000 years: in all, the Earth's history had by now occupied some 75,000 years, and a further 93,000 remained before life would be extinguished by cold. For the first epoch, he revived his theory about a near-collision between the Sun and a passing comet; in the second, he supposed the Earth to have solidified, with the greater part of the fusible rocks on the surface; during the third, all the continents were covered with water; fourthly, the oceans withdrew, and volcanoes built up the land; during the fifth epoch, tropical animals were spread across the whole Earth; in the sixth, the different continents separated; and finally, in the seventh and last epoch, one reached the period of Man's existence.

Taken separately, none of Buffon's chief steps was entirely new. Descartes had spoken of the planets as having formed out of incandescent stars, Leibniz had developed this idea further in his *Protogea*, and even Newton had used the same idea to explain the flattening of the Earth at the Poles. (This was the result of centrifugal action while the planet was still hot and plastic.) The idea of interpreting the Days of Creation as geological epochs had been hinted at in the 1690s, during the discussion prompted by Burnet's *Sacred History of the Earth*. Again, Benoit de Maillet's posthumous dialogue, *Telliamed*, had foreshadowed Buffon's extension of the time-scale: the 'Indian sage' into whose mouth de Maillet put his own original speculations had insisted that we should not 'fix a beginning to that which perhaps never had one. Let us not measure the past duration of the world by that of our own years.' Finally, Buffon's experiments on cooling were consciously modelled on those by which Newton had established the basic laws governing the rate at which bodies lose heat. In fact, Newton had gone so far as to enquire, in an incidental note, how long it would take for an iron sphere the size of the Earth to cool down from red heat, and had arrived at an answer comparable to Buffon's—*viz.*: 50,000 years. For once, Newton's calculations led to a result that he could not square with his religious convictions, and he unhesitatingly rejected it: there must be something wrong with his calculation—perhaps his assumption that a large sphere would cool very much more slowly than a small one—and he added: 'I should be glad that the true ratio was investigated by experiments.'

What distinguished Buffon's *Epochs of Nature* was the cumulative weight of his whole argument. He drew together into a unified whole half a dozen ideas which had previously been thrown out independently. Furthermore, he patiently settled down to work out, in numerical terms, the actual periods of time demanded by a physical theory of the Earth's development. To summarize his 1774 figures for the Earth's cooling:

> If we suppose, as all the phenomena seem to indicate, that the earth was once in a state of liquefaction caused by fire, our experiments then prove that, if the globe was entirely made of iron or of ferrous matter, it would have solidified as far as the centre only after 4,026 years, cooled to a point at which it could be touched without burning after 46,991 years; and that it would not have cooled to the present temperature until after 100,696 years. But since the earth, so far as we know it, seems to be made up of fusible and calcareous materials which cool in a shorter time than ferrous materials, it is necessary, so as to get as close as possible to the truth, to allow for the respective times of cooling of these different materials, such as we have found them to be in our experiments . . . [So] one finds that the terrestrial globe will have solidified to the centre in 2,905 years approximately, cooled to the point at which one could touch it in 33,911 years approximately, and to the present temperature in 74,047 years approximately.

This last figure he adjusted to 74,832 years, to allow for the effects of the Sun's radiant heat. Using similar arguments, he worked out how long all the different planets and their satellites must have taken to reach habitable temperatures (see table on page 148).

These calculations may have been over-simplified, but they were a beginning. Newton could not explain how the globe became habitable after no more than a few centuries, as the Biblical time-scale required. Buffon replied that it could not possibly have done so, short of an arbitrary and unreasonable supernatural intervention. No doubt his own figures were far greater than those commonly believed to possess Biblical authority, but one must not fly in the face of reason. Men were no longer intimidated by the vast distances of Space, nor did they find it difficult to imagine (e.g.) one hundred thousand pounds in money: why, then, should their minds recoil from the idea of seventy or a hundred thousand *years*? In fact, the figures he calculated were the absolute minimum required for the

Beginning, end and duration of existence of living organisms in each planet
[Buffon, 1774]

[*N.B.: Origin of the planets assumed to date back* 74,832 *years, i.e. to* 73,058 B.C.]	*Beginning* (*from formation of planets*)	*End*	*Overall Duration*	*Years still to run*
5th satellite of Saturn	5,161	47,558	42,389	[Extinct for 27,274]
MOON	7,890	72,514	64,624	[Extinct for 2,318]
MARS	13,685	60,326	56,641	[Extinct for 14,506]
4th satellite of Saturn	18,399	76,525	58,126	1,693
4th satellite of Jupiter	23,730	98,696	74,966	23,864
MERCURY	26,053	187,765	161,712	112,933
EARTH	35,983	168,123	132,140	93,291
3rd satellite of Saturn	37,672	156,658	118,986	81,826
2nd satellite of Saturn	40,373	167,928	127,655	93,096
1st satellite of Saturn	42,021	174,784	132,763	99,952
VENUS	44,067	228,540	184,473	153,708
Saturn's ring	56,396	177,568	121,172	102,736
3rd satellite of Jupiter	59,483	247,401	187,918	172,569
SATURN	62,906	262,020	199,114	187,188
2nd satellite of Jupiter	64,496	271,098	206,602	196,266
1st satellite of Jupiter	74,724	311,973	237,249	237,141
JUPITER	115,623	485,121	367,498	[not yet begun]

formation of the Earth; and, if they were hard to grasp, we must simply remind ourselves that

> there is no difference [in authority] between the truths that God has revealed [in scripture], and those which He has permitted us to discover by observation and enquiry.

In one respect, Buffon remained a typical eighteenth-century natural philosopher. For all the modernity of his tone and his arguments, he took it for granted that 'whatever can exist, does exist': it was enough to calculate when each of the bodies in the solar system would reach a temperature capable of supporting life, and thereupon life had presumably appeared. Like Fontenelle before him, he treated the plurality of worlds as implying a plurality of *inhabited* worlds:

> (i) Organized nature as we know it, is not yet born on Jupiter, whose heat is still too great today for one to touch its surface, and it will only be in 40,791 years [i.e. 115,623–74,832: cf. table above] that living creatures will be able to subsist there, but thereafter once established they would last 367,498 years on that large planet;

(ii) living nature, as we know it, has been extinct on the fifth satellite of Saturn for the last 27,274 years; on Mars, for the last 14,506 years, and on the Moon for the last 2,318 years; [etc. etc.] . . .

[Hence my belief in] the real existence of organized and sensible beings on all the bodies of the solar system, and the more-than-likely existence of the same beings on all the other bodies making up the systems of other suns, so augmenting and multiplying almost to infinity the extent of living Nature, and at the same time raising the greatest of all monuments to the glory of the Creator.

By our standards, of course, Buffon's calculations had given the Earth not too long, but far too short a life. Where did his physics go astray? He acknowledged that the present temperature of any planet or satellite must be determined by two separate factors—the radiant heat falling on it from the Sun, and the residual heat remaining from its original molten state—and he did his best to estimate the relative contributions of these two factors. In the 1770s the laws of normal cooling were well established, but radiant heat was little understood: Buffon mistakenly decided that the effects of solar radiation were of secondary importance as compared with the Earth's residual heat. He would have reached a more accurate result if he had assumed the exact reverse. The present temperature of a planet or satellite depends almost entirely on the amount of incoming solar radiation: to a first approximation, one can neglect the heat coming from the interior of the planet itself. This fact became evident soon after 1800, when physicists began to study radiant heat seriously, and within forty years of their original publication Buffon's calculations were completely undercut by the new ones of Fourier.

Scientifically, then, Buffon's pioneer attempt to estimate the age of the Earth by appeal to physical principles had been a failure. His religious compromise, equally, had satisfied none of the parties affected. So, in retrospect, the *Epochs of Nature* may seem at best a magnificent ruin. Yet, this verdict would be unjust. Buffon's figures were wrong; but his was the voice of the future, and—above all—his books were very widely read. Moreover, his calculations had proved the essential point: that the time-barrier could be breached. By invoking the laws governing familiar physical processes, such as cooling, one might infer the former state of things from the present face of Nature, and determine the dates of physical events far earlier than the first human records. If Buffon had lived to see the next fifty years of geology, he would have been well content, for by the 1820s literal-minded fundamentalism was in full retreat, and the defenders of orthodoxy were thankful to take refuge in his own interpretation of the Days of Creation. The details of his argument could go overboard: he had made the points that mattered.

As things turned out, Buffon's own theories about the age of the Earth made little immediate contribution to geology proper, and the cosmological approach to the subject soon went out of fashion. For his was a speculative and roundabout way of arguing, in which the history of the Earth was deduced indirectly from a general theory of the planetary system, rather than being pieced together directly from the actual evidence of geological exploration. The results might carry a certain abstract conviction to Newtonian natural philosophers, but they had no great relevance to the experience of practical men, and they were not easily reconciled with the ideas such men inherited from their predecessors. Indeed, as we shall see later, until well into the twentieth century certain glaring discrepancies remained between physical cosmology on the one hand, and zoology and geology on the other.

Meanwhile, however, other men were scrutinizing the Earth's surface directly, and beginning to enquire about the agencies that had shaped it into its present form. At first—as we said—they were not moved by historical curiosity: they were concerned merely to map the rock-strata found in the Earth's crust at different places, and to discover whether there was any common sequence in the formations overlaying one another in different countries and regions. It was some time before the discovery of a widespread *geographical* order in the nature of the crust came to be recognized as evidence of a *temporal* order in the processes by which the rocks had come into existence—instead of being accepted unquestioningly as the pattern imposed at the original Creation. And it took most of a century for geologists to establish what agencies had been involved in these processes of formation, to discover how one might compare the ages of strata geographically distant from one another, and to build up a consistent history of the Earth's crust from the evidence of its present form and fossil content.

To begin with, there were two chief centres of geological research, one in Germany, the other in France. Ever since mediaeval times, German craftsmen had been building up a tradition of mineralogy and mining, and this practical aspect marked the first German contributions to the new science. Anyone with first-hand experience of mining technique knew something about the stratification of rock-layers, and much was done to lay the foundations for geology by simply describing and naming the different types of rock and rock-strata. The chief leader in this work was the Saxon geologist, A. G. Werner. Like Boerhaave, Werner was one of the supreme scientific teachers, who made his mark primarily by stimulating the interest of his students; and as with that other great teacher,

Linnaeus, the men who learned from him were won over by his intimate and detailed mastery over his subject-matter, rather than because of his theoretical penetration. In geology as much as in zoology and botany, what the eighteenth century demanded was a comprehensive and orderly classification, together with a precise nomenclature. These Werner provided: he classified the superimposed rock-strata into four or five main types, and several dozen subdivisions—ranging from granite and gneiss, in which fossils were never found, up through the various fossil-bearing strata to the sands, clays and volcanic lava of the surface layers—and he saw that the superposition of layers must have a historical significance also, the fossil-bearing rocks being younger than the granite, and the superficial rocks being the most recent of all.

In France, meanwhile, other historical clues were coming to light. In 1751, J. E. Guettard stopped at Montelimar on his way back from a journey throughout southern Italy. His attention was caught by the fact that the paving of the streets was of a type of stone—hexagonal basalt—strikingly like some which he had seen in the volcanic regions around Vesuvius, and that the milestones and even some of the local buildings were also made of volcanic-looking stone. He enquired where this stone had come from, and was directed into the mountains west of the Rhône. Here he found a region of steep river valleys, containing dramatic cliffs of basalt 'organ-pipes', and leading to a central plateau dominated by a range of conical peaks. One of these mountains was the Puy de Dôme, the site of Pascal's experiments on atmospheric pressure. In the neighbouring village of Volvic, men had been quarrying a durable black stone for centuries. It had been used to build the cathedral at Clermont Ferrand, and is still used for milestones on French roads to this day. Until Guettard arrived, however, the quarries of Volvic and the neighbouring peaks had kept their most important secret, which could be discerned only by men who looked at them with the right questions in mind. As Guettard immediately realized, the surrounding mountain peaks were the cones of extinct volcanoes, and the paths of their ancient lava-streams could be clearly seen in the surrounding countryside. One had only to strip off the surface disguise of top-soil and vegetation, and the whole region was recognizable as a vast area shaped by volcanic action.

Like the first deciphering of a hitherto-unintelligible script, Guettard's discovery precipitated a chain of others. Once the first step was taken, everything else fell neatly into place. During the decades that followed, similar regions of extinct volcanoes were recognized in a dozen parts of the world, many of them associated with the hexagonal basalt characteristic of Giants' Causeways. Yet the puzzle remained: if the time-span of the world was really less than 6000 years, how could such violent volcanic action have gone on unrecorded? It seemed impossible that all these formidable eruptions could have taken place during the short span of time before the earliest human records. Faced with evidence such as this men were gradually driven towards the conclusion which they had so long resisted: that the present phase in the Earth's history was only the most recent in a series of prolonged epochs, and that the face of the globe had changed radically from age to age.

The Agents of Geological Change

Though as late as 1790 the fact of geological change was still not universally conceded, there were many geologists who recognized the force of the evidence. But these geologists were themselves divided about the agencies responsible, and two main schools began to form. It had long been known that both heat and water were actively changing the face of the Earth during the present era, and presumably had done so in former ages also. Opinion was now sharply divided over the question which of these two agents had played the dominant part. At this point, Werner—perhaps unfortunately—departed from his own empirical principles. Though he was primarily a geological observer, his ideas had been strongly influenced by the Biblical tradition of the Flood, and this fact governed his own interpretation of geological facts. Rock-strata had been universally formed by 'sedimentation': minute particles being deposited on the bed of the ocean in successive layers, exposed by the recession of the waters, and finally weathered and dissected by the action of rain and rivers. Werner convinced himself that, historically, volcanoes were a comparatively recent phenomenon so that, geologically, volcanic rocks should be no more than superficial features. Despite the frequent occurence of basalt in volcanic regions, he felt bound to deny that this was any indication of its origin: for, in many places, basalt occurs quite deep in the Earth, and an 'igneous' origin for all basalt would imply that the period of volcanic action stretched back to an epoch when (as he believed) the whole surface of the Earth was submerged below a universal ocean.

With the dispersion of Werner's students throughout Europe, his theoretical views acquired a certain orthodoxy: this 'Neptunism' could apparently explain the origin of most rock-strata, and Werner found intellectual expedients for disposing of the embarrassing remainder. The theory had the further advantage of offering to reconcile Biblical and geological history: Werner's ocean was identified with Noah's Flood. There were, of course, plenty of unsolved problems: where, for example, had the surplus water disappeared, after retreating from the existing continents? The problem of the time-scale, too, was a continual embarrassment, for sedimentation is undeniably a very slow process. But the more severe these other difficulties became, the more urgent it appeared to hold fast to the one point of reconciliation. As late as 1825 many geologists, in the hope of maintaining a harmony between *Genesis* and geology, were still searching for Neptunist explanations even of those formations which were much more readily explicable as the products of heat.

The 'Vulcanist' geologists, who allowed heat a large part in the shaping of the Earth at all epochs, were accordingly exposed not just to scientific criticism but to contumely. (It was perhaps unfortunate that the Bible mentioned a Flood, for volcanic action is much quicker than sedimentation and Vulcanism did not suffer the same embarrassments over the time-scale.) Too much was in fact at stake for the Neptunists to admit any rival, co-equal agency. So, when Goethe alluded to this controversy in Part II of *Faust*, he significantly made Mephistopheles the spokesman for the igneous theory—

> The Devils all set up a coughing, sneezing:
> At every vent without cessation wheezing:
> With sulphur stench and acids Hell dilated,
> And such enormous gas was thence created,
> That very soon Earth's level, far extended,
> Thick as it was, was heaved, and split and rended!

Faust was content to regard the mountain-ranges as part of Nature's original creation, and looked to geology only to explain the subsequent cutting-away of the valleys:

> To me are mountain-masses grandly dumb:
> I ask not Whence? and ask not Why? they come.
> When nature in herself her being founded,
> Complete and perfect when the globe she rounded,
> Glad of the summits and the gorges deep,
> Set rock to rock, and mountain steep to steep,
> The hills with easy outlines downward moulded,
> Till gently from their feet the vales unfolded!
> They green and grow: with joy therein she ranges,
> Requiring no insane, convulsive changes.

The ideas of one man, in particular, suffered as a result of these emotionally-based attacks. Whereas Werner spent practically his whole life in Saxony, James Hutton was a more widely-travelled man. He understood the geological effects of the seas, rivers and rain very well, but he could not agree that these had played the dominant part in shaping the Earth. The effects of water were almost entirely destructive: rocks at a higher level being broken down and carried away, to be deposited lower down, in alluvial plains, in deltas or on the sea-bed. But the existence of mountain-ranges and other uplifted features of the Earth's crust called for another, more constructive agency; and, being a pupil of James Black at Edinburgh, he saw how subterranean heat could cause surface rocks to expand, buckle and be thrust upwards. To us, 'Vulcanism' or 'Plutonism' in this form appears a natural complement to Neptunism, and that (as we shall see) is what Hutton intended it to be. In the atmosphere of the time, however, his *Theory of the Earth* appeared not so much to complete Neptunism as to deny it, and he found himself faced by a far greater weight of opposition than he was really entitled to expect.

The bitterness and virulence aroused in Hutton's opponents is intelligible today only if one recalls the ideological situation created by the French Revolution. Even before 1789, the atmosphere had been unfavourable to anything radical in the way of geological speculation, and the watchdogs of respectability were already defending what Joseph Townsend later called 'the character of Moses for veracity as a historian'. The aftermath of the French Revolution added a spur to their anxieties: in some way or other, they reasoned, the terrors and brutalities afflicting France in the 1790s must be results of the impiety and free thought common among French intellectuals earlier. For the next generation, the intellectual debate lost its cool, classical eighteenth-century tones, and took on new, more hysterical, overtones. The French Revolution was one of the great traumas of history, shaking the optimistic faith of eighteenth-century thinkers as completely as the Reformation had shaken that of sixteenth-century Catholics and the Russian Revolution was to shake twentieth-century liberal democrats. Once again, science was caught up with ideology, and intellectual doctrines were criticized not so much for their scientific adequacy as for their supposed moral 'tendencies'.

James Hutton himself came under the lash of Richard Kirwan who, though a perfectly competent chemist, was moved to defend Biblical history, word for word, against Hutton's own view. (In the same way Chateaubriand, in his *Génie du Christianisme*, pitted Buffon against Moses.) The new bitterness is clear from Kirwan's words:

> Recent experience has shown that the obscurity in which the philosophical knowledge of [the original] state [of the Earth] has hitherto been involved, has proved too favourable to the structure of various systems of atheism or infidelity, as these have been in their turn to turbulence and immorality.

Hutton's very caution and moderation themselves caused offence. If he had attacked the fundamentalist view directly or, like Buffon, had forcibly adapted it to suit the geological evidence, there would at any rate have been something for the orthodox to criticize. His actual procedure struck them as even more insulting: he simply *ignored* the sacred tradition, and his silence about Moses and the Flood, though originating in modesty, was interpreted as scorn.

The Perspective of Indefinite Time

It was all very unjust. James Hutton's handling of the geological evidence might seem revolutionary, but his fundamental aims were conservative and devout. Philosophically, he was as close in spirit to Isaac Newton a century earlier as he was to his successor of forty years later, Charles Lyell. Certainly, Hutton refused to shut his eyes to the historical implications of the new geology, and sheer intellectual honesty compelled him to declare that it revealed no trace of the world's original Creation; yet, in most other respects, his theoretical system tended rather to confirm and complete the orthodox Protestant conception of Nature, as the orderly product of divine wisdom. Where Newton had explained the providential stability of the solar system as a balance between gravitational and centrifugal tendencies, Hutton now saw the Earth's structure as maintained in a habitable state throughout indefinite time by a similar providential balance of geological agencies.

Hutton took a clear-eyed view of the evidence which geologists had brought to light. Man was evidently a somewhat recent inhabitant of the globe, and all the evidence suggested that the Earth was of very great antiquity, having passed through a sequence of lengthy epochs earlier than our own. About its state during these earlier epochs, we could draw only limited conclusions: there was no reason to suppose that, from geological features visible today, we could reason back through time more than part-way towards the Earth's origin. Nor could one draw a sharp line (as both Buffon and Werner had done) dividing a more recent, volcanic epoch from an earlier, aqueous one, during which water had been the dominant agent. On the contrary, the evidence was that all the geological forces active today had been active in every previous epoch also. Furthermore, all these forces acted very slowly. The changes a man could actually observe for himself within his own lifetime were only a minute sample of those which the same agencies might bring about—given indefinite time: he saw, as it were, only a single 'still' from the whole moving picture of geological change. Yet, given time, one could perfectly well envisage these forces producing all the geographical features with which we are familiar:

> Great things are not understood without the analysing of many operations, and the combination of time with many events happening in succession.

Here was the germ of Lyell's 'uniformitarian' method. If time alone were all the concession the evidence demanded, surely we had no adequate reasons for refusing this, or for limiting the number of years over which Nature was supposed to have acted. The geographical effects of geological change might be too slow to detect. Even by comparing our own knowledge of the Earth with that of ancient authors, we could not estimate the age of the continents—

> It is in vain to attempt to measure a quantity which escapes our notice, and which history cannot ascertain; and we might just as well attempt to measure the distance of the stars without a parallax, as to calculate the destruction of the solid land without a measure corresponding to the whole.

Yet, however slow it might be, the fact of geological change was undeniable. Once we allowed them sufficient time, fire, heat, weather and water could between them bring into being any geological or geographical feature we chose to consider.

The spectacle of mechanical agencies operating uniformly throughout 'indefinite successions of ages' was a forward-looking one, yet in Hutton's own person it was associated with a thoroughly devout general philosophy. His contemporaries mistakenly accused him of importing into Scotland the wanton French atheism of Baron d'Holbach's *System of Nature*. Yet there was, in fact, nothing in Hutton's system—apart from the unbounded chronology—that could legitimately give offence. In his own eyes, he shared John Ray's conviction that the structure of the world displayed the wisdom of its Author. All the geological changes which had gone to form the present continents had served to create a habitation fit for Man. The great lapse of time involved should not deceive us into thinking of Nature's operations as random and aimless:

> Though, in generalizing the operations of nature, we have arrived at those great events, which, at first sight, may fill the mind with wonder and with doubt, we are not to suppose, that there is any violent exertion of power, such as is required in order to produce a great event in little time; in nature, we find no deficiency in respect of time, nor any limitation with regard to power. But time is not made to flow in vain; nor does there ever appear the exertion superfluous to power, or the manifestation of design, not calculated in wisdom to effect some general end.

The Earth was something more than a chance collection of material objects acted on by physical forces. It was demonstrably a 'system'—one as intricate, well balanced and self-adjusting as the planetary system itself.

> In order to understand the system of the heavens, it is necessary to connect together periods of measured time, and the distinguished places of revolving bodies. It is thus that system may be observed, or wisdom, in the proper adapting of powers to an intention. In like manner, we cannot understand the system of the globe, without seeing that progress of things which is brought about in time, thus measuring the natural operations of the earth with those of the heavens.

The counter-balancing actions of heat and water, which were continually building up and breaking down different parts of the Earth's crust in such a way as to maintain it in a form fit for animal and human life, provided fresh evidence of the Creator's foresight and design. True: geologists were in no position to reconstruct the whole back-history of the Earth. But this (Hutton concluded) did not seriously matter. It was enough to demonstrate how the systematic interaction of geological processes played its proper part in the providential scheme:

> We have now got to the end of our reasoning; we have no data further to conclude immediately from that which actually is: But we have got enough; we have the satisfaction to find, that in nature there is wisdom, system, and consistency. For having, in the natural history of this earth, seen a succession of worlds, we may from this conclude that there is a system in nature; in like manner as, from seeing revolutions of the planets, it is concluded, that there is a system by which they are intended to continue those revolutions. But if the succession of worlds is established in the system of nature, it is in vain to look for any thing higher in the origin of the earth. The result, therefore, of our present enquiry is, that we find no vestige of a beginning,—no prospect of an end.

Starting from quite different assumptions, Buffon and Hutton had reached the same conclusion: that the Earth had existed very much longer than had customarily been supposed. For Buffon, this conclusion followed from an argument about the origin of the Earth. Hutton set aside all questions about this origin, but arrived at the same conclusion from a study of continuing geological processes. He stated his position very moderately, and took care both to avoid exaggerating his claims and to respect orthodox prejudices in natural theology. Thus, he never denied that there had been an original Creation: he insisted only that the geology of his time revealed 'no vestige of a beginning', providing evidence rather of 'a time indefinite in length'. (Here, too, Hutton was only completing the picture of Nature begun by Isaac Newton. In the Newtonian account of the planetary system, the stable orbits of the planets were established by divine intervention, and their present forms proved nothing about the process of Creation itself. Likewise, for Hutton, the present geological order proved nothing about the origin of the Earth.)

If Hutton had had less of a taste for Protestant natural theology, he might have carried his argument one stage further. As things were, he ended, like Newton, in the philosophical position of Plato's *Timaeus*—the existence of a stable and rational order in Nature being explained as the consequence of a hypothetical Creation, at some indefinite moment of the remote past. Yet this was not the only possible interpretation of his theories. He might with equal consistency have extrapolated his argument, dispensed with the hypothetical Creation entirely, and used his discoveries to argue that past time was not merely indefinite but infinite. This was the position taken up in antiquity by Aristotle in order to avoid the embarrassments arising out of Plato's theory. Throughout the Middle Ages and the Renaissance, the Aristotelian doctrine had been a heresy for Catholics and Protestants alike; and Hutton, too, now held back from taking this last step. That was left to a minor admirer of his, a physician called George Hoggart Toulmin, in a series of treatises on the antiquity, duration and eternity of the world published from 1780 on—immediately before Hutton's fundamental paper.

The cautious Hutton had found in geology only an absence of evidence about the Earth's original beginning and eventful end: Toulmin, more positively, read the geological evidence for the Earth's stability as proof that Nature was in fact eternal:

> A succession of events, something similar to what is continually observed, has *ever* taken place; nature having, through an eternal period of duration, acted by laws fixed and immutable. In the extensive circle of existence, therefore, in vain do we seek for the beginning of things. How fruitless every recourse to calculation on the subject of the stars and nature's first existence! The stretch of human conception necessarily fails us, multiplied series of numbers, of which we cannot possibly have any adequate idea, unavoidably stop short and leave the matter removed at an unlimited distance.

A great many facts indicated the antiquity of the Earth, the animals, the human race and social life equally; and, like Aristotle, Toulmin could find no reason to restrict the power of Nature by limiting the time available for her action. If the laws of nature were fixed, it was only reasonable to conclude

> that the human species, the world, the suns, the planets, and the sublime intelligent beings and objects of the universe, having never had any beginning, have existed with their various modifications uncaused through all eternity.

This Aristotelian view was, indeed, the logical destination towards which orthodox eighteenth-century ideas about the Order of Nature had been pointing all along.

The geological evidence available in the 1790s could thus be used to justify either Hutton's version of the 'Big Bang' myth of the *Timaeus* or alternatively Toulmin's revival of the Aristotelian 'Steady State'. (It could, for that matter, have been used also to support a neo-Stoic Cyclical Theory of the Earth—one in which fire and water took the upper hand by turns, in an endless sequence of geological phases.) This was no accident. For, as Hutton rightly conceded, the geological evidence gave by itself only a restricted view of the past—bounded within a period of time throughout which, for all that one could tell, the forces of Nature had been operating in the same, constant way. The historical time-barrier of earlier centuries was merely being exchanged for a geological time-barrier farther back into the past. As to the period before that, geology left one as much in the dark as ever. The new science explained only the agencies responsible for shaping particular geographical features or rock-formations: the question, whether the Earth-as-a-whole had or had not been created at some still earlier moment of time, was beyond its scope. For lack of further solid evidence and arguments, that question could be answered only in terms of *a priori* speculations; and at this level Newton and Hutton could make no real advance on Plato, nor Toulmin on Aristotle. Even Buffon's account of the matter, for all its veneer of arithmetical detail, was as much a piece of neo-Ionian theorizing as a serious essay in physical science. The choice between an initial Moment of Creation, an Eternal Order, a Cyclical Cosmos and a continuing One-Way Process remained what it had always been: not an empirical conclusion which could be tested against the facts of observation, but rather a prior philosophical decision governing one's interpretation of those facts.

The Historical Time-Barrier is Broken

After the theological thunder of the 1790s had subsided, geologists mostly preferred discretion to valour. For the next twenty years, they limited the scope of their speculations, and concentrated on consolidating the empirical foundations of their science. This was a healthy reaction. There are times in the development of any science when philosophical debate has done all it can to clarify the fundamental questions, and a full generation's work must intervene before anyone can hope to give adequately-founded answers. This had been the position in geology when Werner first condemned premature theorizing, and called on his pupils to go out—with muddy boots and earthy hands—and make a detailed, first-hand study of the rock-strata now forming the Earth's crust. It was still the position in 1800; for the whole argument between fire and water, Neptunists and Vulcanists, had gone off at half-cock.

Yet the fundamentalists were right to smell danger, for the crucial battle over chronology was about to be joined. In Book VIII of Milton's *Paradise Lost*, the Archangel Raphael had turned aside Adam's questions about cosmology, advising him not to pry too closely into what was God's proper concern. Now, in Book III of his poem *The Task* (1785), William Cowper commented with some asperity on the renewed curiosity about the origin of the world—

> Some write a narrative of wars, and feats
> Of heroes little known; and call the rant
> An history: . . . Some drill and bore
> The solid earth, and from the strata there
> Extract a register, by which we learn
> That he who made it, and reveal'd its date
> To Moses, was mistaken in its age.
> Some, more acute, and more industrious still,
> Contrive creation; travel nature up
> To the sharp peak of her sublimest height,
> And tell us whence the stars; why some are fix'd
> And planetary some; what gave them first
> Rotation, from what fountain flow'd their light.
> Great contest follows, and much learned dust
> Involves the combatants; each claiming truth
> And truth disclaiming both. And thus they spend
> The little wick of life's poor shallow lamp,
> In playing tricks with nature, giving laws
> To distant worlds, and trifling in their own.

There was no necessity for scientific understanding to undermine a sense of wonder at the divine Creation: unfortunately (as Cowper saw it) scientists were wantonly precipitating a conflict between science and religion—

> God never meant that man should scale the heavn's
> By strides of human wisdom. In his works
> So wondrous, he commands us in his word

To seek *him* rather, where his mercy shines . . .
But never yet did philosophic tube,
That brings the planets home into the eye
Of observation, and discovers, else
Not visible, his family of worlds,
Discover him that rules them; such a veil
Hangs over mortal eyes, blind from the birth,
And dark in things divine. Full often, too,
Our wayward intellect, the more we learn
Of nature, overlooks her author more;
From instrumental causes proud to draw
Conclusions retrograde, and mad mistake.

It was time for a truce to the theoretical debate. When the Geological Society of London was established in 1807, it resolved as a deliberate policy to take no sides in arguments about the Theory of the Earth, and to accept meritorious contributions from geologists of any theoretical persuasion or none. The bitter disputations of Edinburgh were abandoned in favour of more pedestrian—but more productive—lines of enquiry.

All theories apart, Hutton and Werner had each helped to point the way along which geology would advance: Werner, by his insistence on first-hand field study and accurate classification, Hutton by his 'uniformitarian' method of argument. For scientists (as Playfair declared, in his *Illustrations of the Huttonian Theory*) could interpret the processes of geological change satisfactorily only by comparing them with other natural phenomena which they could observe at the present time:

> If it is once settled, that a theory of the earth ought to have no other aim but to discover the laws that regulate the changes on the surface, or in the interior of the globe, the subject is brought within the sphere either of observation or analogy; and there is no reason to suppose, that man, who has numbered the stars, and measured their forces, shall ultimately prove unequal to this investigation.

These maxims of method were given fresh force by the work of Sir James Hall. Hutton himself feared that the processes responsible for the formation of strata might be too violent to reproduce in the laboratory. Hall set this fear at rest and, in doing so, settled one thorny issue in Hutton's favour. The geographical association between basalt platforms (Giants' Causeways) and active or extinct volcanoes had been a commonplace since Guettard's time. Yet no active volcano is ever seen to erupt streams of basalt columns, nor does lava left to cool on the Earth's surface ever crystallize out into hexagonal form; so Hutton's theory that basalt was formed by subterraneous volcanic action had remained unproved. Now, however, James Hall succeeded in reproducing the physico-chemical conditions encountered by a stream of lava when forced into a line of weakness between two subterranean strata—intense heat and enormous pressure, without any outlet for the escape of gases. If cooled very slowly under such condition, molten minerals did indeed solidify into regular crystals, rather than into the formless slag or tufa familiar from lava-streams on the surface. So geology became an experimental, as well as an observational science: by actually demonstrating the novel mechanisms invoked in their theories, geologists strengthened their intellectual bridges back into the past.

Stratigraphy

Meanwhile, an even more significant method of enquiry was being developed by a man who, at first, was not recognized as being a geologist at all. William Smith was a self-made craftsman, the son of a blacksmith in the Cotswold village of Churchill. Having served as apprentice to a surveyor at Stow-on-the-Wold, he set up in practice at Bath, being employed to assess the resources of coal mines and to construct the Somerset Coal Canal. (This was the period of 'canal-mania'.) His work took him all over England, and he became widely known as a practical engineer; yet all the while, from childhood on, his eyes and his mind were fascinated by the problems of geology. Whenever duty took him down a mine, into a canal-cutting, or on a cross-country journey, he carefully noted the order and succession of the strata (coal measures, limestone, sandstone, fullers' earth, chalk) and collected the organic fossils embedded in each layer. Travelling to the North in 1794, he found similar sequences of strata recurring again and again right across England. Two years later, he had established that a corresponding strata in any part of the country always contained the same characteristic fossils. By 1799, he was already embarking on the first of those systematic geological maps which culminated in his great five-miles-to-an-inch map of England and Wales (1815).

In a word, William Smith was the inventor of 'stratigraphy'; and this new technique did more than anything else to turn geology into an *historical* science. For the example of Smith's work convinced geologists that the strata in all parts of the Earth's crust belonged in a single common sequence—Cambrian, Ordovician, Silurian, Devonian and so on—and that this sequence was not merely a fact of geography, but reflected the temporal order in which the rocks had been laid down. Only a few members of the whole sequence were present at any one location but, among those that were represented, the lower ones were also the older. The superposed strata were thus so many footholds in past time. Moreover, this common sequence of strata was confirmed by the common sequence of fossil species found within them: in fact, the fossil contents of two strata provided an even better basis for comparing their relative positions—and so their relative ages—than their mineralogical contents. So the succession of rock-layers revealed by Smith's stratigraphical research provided an intellectual ladder reaching back into the remotest depths of the past; while the fossils in each layer served as the rungs of this ladder. In due course, this fact was to pose some difficult questions for zoology: had the Earth's population of living things really differed so strikingly from age to age, or was it merely that some fossils were more durable than others? And, if the population was changing, did this imply a succession of new creations, or were organisms actually *changing their forms*? For the moment, however, the facts were enough, and geologists gratefully set to work completing and elaborating the historical sequence of strata. Admittedly, Smith's discoveries were not completely novel. Steno had been aware of a general association between different fossils and strata, and Werner too had seen the presence or absence of fossils as a possible index of their comparative age. But it was William Smith who demonstrated fully and convincingly how the sequence of fossil forms could be used to 'identify the courses and continuity of the Strata in their order of superposition'. As a result, 'palaeontological' studies took a fundamental place in geology, and it at last became possible to estimate the comparative antiquity of formations in different regions or countries.

By the 1820s, geology was acquiring the internal coherence and complexity of a developed natural science. From now on, geologists spent less of their time looking over their shoulders at the Bible, and paid more attention to the internal consistency of their results—to the mechanisms they assumed, and the time required for their operation. This is not to say that they were allowed to forget the religious implications of their science. Smith's friend, Joseph Townsend, had great hopes that the new stratigraphy would confirm Old Testament teaching. But it was no longer necessary for geologists to find the whole Pentateuch rewritten in the rocks, chapter by chapter and verse by verse.

The Doctrine of Catastrophes

The end-result was inevitable. Bit by bit, a picture of the Earth's history was constructed which owed everything to 'the testimony of things', and nothing to the five books of Moses. Before 1780, scientific excursions into the past had been isolated, and open to attack piecemeal; but after 1800 they reinforced one another, so that a cumulative pressure built up. Like plants trapped beneath a layer of asphalt, the new geological discoveries pressed more and more forcibly against the restraints of the Old Testament chronology, feeling out its weak points and destroying its consistency. At the beginning of the nineteenth century, only a few natural-born heretics like J. B. Lamarck could exclaim

> Oh! how great is the antiquity of the terrestrial globe! and how little the ideas of those who attribute to the globe an existence of six thousand and a few hundred years duration from its origin to the present!
>
> The natural philosopher and the geologist see things much differently in this respect; [given] the nature of fossils spread in such great numbers in all the parts of the exposed globe [and] the number and disposition of the beds, as well as the nature and order of the materials composing the external crust of the globe . . . —how many occasions they have to be convinced that the antiquity of this same globe is so great that it is absolutely outside the power of man to appreciate it in any manner!

But within a couple of decades even the orthodox were having to admit the need for a vast extension of time. Thus Lamarck's rival, Georges Cuvier, though insisting that the Biblical Deluge 'cannot date back much more than five or six thousand years', none the less required the time-scale of antediluvian history to be expanded indefinitely:

> Genius and science have burst the limits of space; and a few observations, explained by just reasoning, have unveiled the mechanism of the universe. Would it not also be glorious for man to burst the limits of time, and by a few observations, to ascertain the history of this world, and the series of events which preceded the birth of the human race?
>
> The astronomers have doubtless gone faster than the naturalists, and the present phase in the theory of the earth is rather like that when some philosophers thought the sky was made of blocks of stone, and the moon the size of the Peloponnese; but, after men like Anaxagoras, there came the Copernicuses and Keplers who opened up the way for Newton; and why should natural history not have its own Newton one day?

Yet for twenty years it still seemed that a satisfactory agreement could be achieved between the testimony of geology and the scriptures, and a last rearguard action was fought, under the leadership of Cuvier in France and Buckland in England. These men re-aligned the two accounts of pre-history, identifying Noah's Flood with the last great submergence which had left its mark on the strata of Western Europe. As Buckland optimistically put it—

> The grand fact of an universal deluge at no very remote period is proved on grounds so decisive and incontrovertible, that had we never heard of such an event from Scripture or any other Authority, Geology of itself must have called in the assistance of some such catastrophe to explain the phenomena of diluvial action which are universally presented to us, and which are unintelligible without recourse to a deluge exerting its ravages at a period not more ancient than that announced in the Book of Genesis.

Cuvier's expansion of antediluvian time gave geology a great deal more freedom of action; but he in turn became entranced by his own Biblical compromise, and was tempted into premature theorizing. He had mapped the geological strata of the Paris region in great detail, and observed—quite correctly—that the superimposed rock-formation of this area showed sharp discontinuities: moving vertically through a section of the terrain, one passed abruptly from strata containing marine fossils to others containing fresh-water fossils, and then, with equal suddenness, into others containing no fossils at all. Turning to other geological reports, he found similar results: nowhere (it seemed) did the Earth's crust display the sort of steady gradation which Hutton's 'uniformitarian' geology might lead one to expect. These geological discontinuities were quite authentic, and for fifty years they posed a real problem. (Charles Darwin, for instance, frequently referred to them as the 'imperfections of the geological record'.) From our own standpoint, the explanation is not hard to see: however uniformly and continuously the agencies of geological change may act, their effects combine to build up new strata at any point only rarely, and these limited periods of geological construction represent between them only a fraction of the Earth's whole history. A section of the crust at any one point corresponds geologically to—say—an Egyptian chronicle containing only a few years at a time, taken at random from different dynasties.

Cuvier was deeply impressed by his discovery. The discontinuities were, in his view, something more than chance by-products of a continuous process. Such abrupt effects must have resulted from equally abrupt causes—convulsive transformations of the Earth's crust, by which whole continents had been engulfed in ice or plunged below the sea:

> This is most easily proved in the case of the last of these catastrophes, which by a double movement flooded and then drained again the present continents, or at least a large part of the earth which forms

> them today. It has left to this day, in the Northern countries, the corpses of vast quadrupeds seized by the ice, and preserved up to our own time with their skin, their fur and their flesh. If they had not been frozen as soon as killed, putrefaction would have decomposed them... Thus one and the same instant killed these animals and glaciated the countries they inhabited. This event was sudden, instantaneous, without any degree, and what can be so clearly demonstrated for this last catastrophe was none the less true for those which preceded it.

The human race had apparently existed only during the last two eras. Deeper rocks gave evidence of prehuman eras characterized by the fossils of simple organisms; and, underlying these rocks again, primitive fossil-free rocks (such as granite) thrust themselves up in mountain-ranges, to form 'a sort of skeleton and rough framework of the earth'. On this view, the history of the Earth comprised a sequence of peaceful eras, during which Nature operated smoothly according to its present laws, separated by violent catastrophes far surpassing anything within the capacity of the present forces of nature. Failing any natural analogy within our experience, the source of these convulsions must be supernatural, and Noah's Flood—whose divine origin was confirmed in the Bible—had evidently been only the last of a long series.

Buckland underlined the edifying lesson of these catastrophes. While the evidence of astronomy exhibited 'the most admirable proofs of design and intelligence originally exerted at the Creation', it nevertheless had the defect of suggesting

> that the system of the Universe is carried on by the force of the laws originally impressed on matter, without the necessity of fresh interference or continued supervision on the part of the Creator.

Geology provided a 'direct and palpable refutation'. God had repeatedly intervened in Nature: the Earth's structure 'is evidently the result of many and violent convulsions subsequent to its original formation'—

> When therefore we perceive that the secondary causes producing these convulsions operated at successive periods, not blindly and at random, but with a direction to beneficial ends, we see at once the proofs of an overruling Intelligence continuing to superintend, direct, modify, and control the operations of the agents which he originally ordained.

Once the latest of the geological catastrophes was identified with the traditional Deluge, Man could be recognized as a creature of the last few thousand years only. If this alone were granted, geologists—like astronomers—

could claim as much time as they pleased for the earlier 'days of Creation'. The focus of the religious issue thus shifted from the History of the Earth to the History of Man, and the violent controversies which were to be triggered off by Darwin's views about the animal ancestry of Man became inescapable.

Uniformitarianism

The catastrophist compromise was too frail to last. It was undercut as early as 1830, when the young Charles Lyell published the first edition of his *Principles of Geology*. This book synthesized thirty years of geological discoveries, demonstrating how—granted enough time—one might account for them all in terms of familiar natural processes alone. Though Lyell had been a pupil of Buckland, his reading of J. B. Lamarck and George Scrope (who was repeating and amplifying Guettard's work on the extinct volcanoes of Central France) converted him to a position more like that of James Hutton: he became convinced that the appeal to catastrophes was quite unnecessary, and that geology must dispense with them. Already, the very number of catastrophes demanded by a detailed study of the geological record (twenty-seven, on some accounts) was becoming an embarrassment: such repeated exercises of the divine Power appeared faintly ridiculous. So Lyell had no need to attack the 'catastrophist' theory head-on. Instead, he made innocent fun of its assumptions, and showed that it was superfluous both for theological and for scientific purposes.

The geological convulsions which Cuvier had invoked to explain the discontinuities between strata were illusory. In the first place, the discontinuities themselves were only *local*:

> We often find, that where an interruption in the consecutive formation in one district is indicated by a sudden transition from one assemblage of fossil species to another, the chasm is filled up, in some other district, by other important groups of strata. The more attentively we study the European continent, the greater we find the extension of the whole series of geological formations. No sooner does the calendar appear to be completed, and the signs of a succession of physical events arranged in chronological order, than we are called upon to intercalate, as it were, some new periods of vast duration.

In the second place, there was a fallacy in the catastrophist argument. Since geological forces were, in the main, quite as much destructive as constructive, one could not expect the geological record to be complete. There must inevitably be long periods of time during which, in any given region, strata were being worn away rather than built up.

> Suppose we have discovered two buried cities at the foot of Vesuvius, immediately superimposed upon each other, with a great mass of tuff and lava intervening, just as Portici and Resina, if now covered with ashes, would overlie Herculaneum. An antiquary might possibly be entitled to infer, from the inscriptions on public edifices, that the inhabitants of the inferior and older town were Greeks, and those of the modern Italians. But he would reason very hastily, if he also concluded, from these data, that there had been a sudden change from the Greek to the Italian language in Campania. . . . So, in geology, if we can assume that it is part of the plan of nature to preserve, in every region of the globe, an unbroken series of monuments to commemorate the vicissitudes of the organic creation, we might infer the sudden extirpation of species, and the simultaneous introduction of others, as often as two formations in contact are found to include dissimilar organic fossils. But we must shut our eyes to the whole economy of the existing causes, aqueous, igneous, and organic, if we fail to perceive that *such is not the plan of Nature*.

In the catastrophist hypothesis, Lyell declared, 'we see the ancient spirit of speculation revived, and a desire manifestly shown to cut, rather than patiently to untie, the Gordian knot'. For his own part, he endeavoured,

> as far as possible, to restrict myself to the known or possible operations of existing causes. . . . The history of the science informs us that this method has always put geologists on the road that leads to truth—suggesting views which, although imperfect at first, have been found capable of improvement, until at last adopted by universal consent.

With this manifesto, he began his reconstruction of geological history, starting from the most recent, tertiary formations, and working gradually down through the strata to the oldest, primary rocks.

Like Hutton forty years earlier, Charles Lyell put aside questions about the origin of the Earth, and limited his enquiries to its development. Whether one could ever hope to discover anything solid about its origin, he was very doubtful: geological evidence and methods of thought threw on light on this, and cosmological arguments (such as Buffon's) rested on questionable assumptions about the sources of the Earth's heat. The only reliable way of arguing back into the past was to consider how the present face of the Earth might have been shaped by the physical agencies we know today. This uniformitarian method was not so much an assumption as an intellectual necessity: what could not be explained in these terms was not really explained at all. So, laboriously and industriously, he set out to show how the formation of all known geological features could plausibly be explained by the action of weathering, volcanic upthrust, sedimentation, erosion and similar processes. If men still found the full potentialities of these simple agencies hard to grasp, that was for two reasons. Their imaginations had not yet expanded to embrace the whole extent of past time; and they were unable to observe for themselves the great majority of constructive geological processes, since these took place either underground or on the sea-bed. Here the position of the human species, as land-animals, was a handicap to science: deep-sea fishes would be in a better position to appreciate the processes of deposition continually building up the ocean floor.

Lyell's position differed from Hutton's in only two serious respects. Firstly, he set less store on the providential character of geological change: he was content to take things as he found them, without feeling the need to convince himself always that they exemplified the Creator's wisdom and foresight. Secondly, where Hutton's account of geological development had inevitably been only schematic, his own could be elaborate and detailed. The intervening forty years had left their mark. By the late 1820s the geological structure of Europe had been very fully mapped—notably, that of the British Isles, Saxony, the Paris Basin, and the volcanic regions of France and Italy—while a good deal had been discovered about other parts of the world also. Lyell therefore had at his disposal a much larger and more varied range of examples, and the range of mechanisms he could illustrate and establish was correspondingly larger and more varied. Instead of the earlier crude opposition between fire and water, he could demonstrate the geological effects of a dozen different agencies, acting either in combination or against one another; and it was the marginal balance between all these agencies, at any one place and time, which determined whether the Earth's crust was being built up or worn down at that point.

Lyell's uniformitarian history of the Earth did not make its way immediately in all quarters. Many people were still too deeply committed to the older traditions, and certain real problems remained unsolved—e.g. the occurrence of those large isolated boulders which are now known to have been transported by former glaciers. (Louis Agassiz formulated his theory of the Ice Ages only in the 1840s.) Yet Lyell had his reward when his old teacher Buckland—without wholly abandoning the attempt to reconcile geology and the scriptures—conceded one of his central points: that the time required to account for the geological development of the Earth far outran the orthodox Biblical chronology. Lyell himself compared this expansion of the time-scale with the seventeenth-century revolution in men's ideas about space—

> It was not until Descartes assumed the indefinite extent of the celestial spaces, and removed the supposed boundaries of the universe, that a just opinion began to be entertained as the relative distances of the heavenly bodies; and until we habituate ourselves to contemplate the possibility of an indefinite lapse of ages having been comprised within each of the more modern periods of the earth's history, we shall be in danger of forming most erroneous and partial views in Geology—

And the rich profusion of geological remains in the *Massif Central* inspired George Scrope to exclaim:

> The leading idea which is present in all our researches, and which accompanies every fresh observation, the sound which to the ear of the student of Nature seems continually echoed in every part of her works, is—
>
> Time!—Time!—Time!

FURTHER READING AND REFERENCES

The broad development of geological ideas is well treated in

C. C. Gillispie: *Genesis and Geology* and *The Edge of Objectivity*
John C. Greene: *The Death of Adam*
Francis C. Haber: *The Age of the World*
Loren Eiseley: *Darwin's Century*

The classical, and more technical, histories of geology are those by Karl von Zittel and Sir Archibald Geikie: there is also much useful material in

F. D. Adams: *The Birth and Development of the Geological Sciences*

There is need for a new, full-length study of James Hutton and his intellectual background. For the moment, one must refer to the essays by Sir E. B. Bailey and S. I. Tomkeieff in the *Proceedings* of the Royal Society of Edinburgh (1947–49). Tomkeieff also discusses the contributions of George Hoggart Toulmin, as a possible source of Hutton's ideas. See also the essay on Hutton by Donald B. McIntyre in *The Fabric of Geology*, ed. C. C. Albritton jr.

Little has been written about William Smith, apart from the short essay

L. R. Cox: *William Smith and the Birth of Stratigraphy*

For an up-to-date discussion of geological dating, see

Patrick M. Hurley: *How Old is the Earth?*

8

The Background to Darwin

THE historical revolution in geology precipitated a similar transformation in zoology, just because the problems of the two sciences, and their evidence, were closely connected. By 1820, palaeontology was becoming a central strand in geological research; and, with Lyell's final breaching of the time-barrier, zoologists too were free to contemplate theoretical possibilities—e.g. that of an evolutionary genealogy—which had not previously been open.

By contrast, the intellectual situation before 1815 had been unfavourable to questions about an origin of species *in time*. Lacking the new geological perspectives, the naturalists of the eighteenth century did not have the scope needed for a comprehensive historical account of the development of organic life. They were preoccupied with very different questions about species from those Darwin asked: questions which were in some ways even more fundamental, having to do, not with the *origin* of species, but with their very *existence*. During the years between 1600 and 1780, in fact, the existence of genuine organic 'species' in Nature was never established beyond all doubt. Many eighteenth-century zoologists questioned their reality: they argued that the classification of living things into distinct groups was an arbitrary operation. A species was an intellectual fiction, not a reality—'Nature knows nothing but individuals'.

If this view had been generally accepted, the world of organisms would have been reduced to an unstructured multitude, with no definite species either to 'originate' or to 'evolve'. To that extent, the demonstration that there are in Nature genuinely distinct kinds of living creature was a real advance; and the eighteenth-century debate about the reality of fixed and definite species, associated with the name of Linnaeus, provided an essential part of the intellectual sub-structure on which the new historical zoology of the nineteenth century was to build.

Our task, at this point, is to sort out the main strands of argument and enquiry that led up to Darwin's work, and so to set the stage for his evolutionary theory. Four groups of problems will claim our attention: the reality of organic species, their geographical distribution, the temporal sequence of fossils, and the mechanisms of inheritance and embryological development. Between 1740 and 1830, much of the substructure for Darwin's *Origin of Species* had been put in place, and Lyell's *Principles of Geology* provided the last indispensable element. Yet, as we shall see, for nearly thirty more years the argument about organic species hung fire, while one last theoretical impasse delayed the historical transformation of zoology.

Are There Fixed Species in Nature?

If one remarks on the similarity between the zoological views of Linnaeus in the 1740s and those of Aristotle 2000 years earlier, one may be tempted to assume that nothing had changed during the intervening time: i.e. that Linnaeus was reproducing uncritically doctrines taken from the earlier scientist. Some people have in fact concluded that the belief in the fixity of species was imposed on European thought very early, as a result of mediaeval scholars' excessive reliance on Aristotle's authority. The truth is that the eighteenth-century zoologists were not just parroting an Aristotelian tradition. Rather, they had reached a standpoint similar to Aristotle's through their own intellectual efforts; and they were, as a result, more firmly convinced of the immutability of organic species than anyone else during the intervening centuries. Certainly the fixity of species was not assumed dogmatically during the Middle Ages. Albert the Great, one of the most influential thirteenth-century scientists, was quite ready to believe that some kinds of living creature (notably, plants) could change into one another and, though many mediaeval scholars were sceptical about these alleged 'changes of species', the possibility of organic transmutation was debated for centuries without bitterness or ridicule. In fact, mediaeval scientists could hardly insist on the fixity of species, since they had no clear criteria for distinguishing one species from another: modern taxonomy, as a disinterested science, dates only from the late sixteenth century. So, for better or for worse, the evidence is clear: the doctrine of absolute fixity took its classic form only *after* the Renaissance.

When in the 1690s John Ray acclaimed the richness of organic nature as evidence of *The Wisdom of God*, he did so as the leading authority on classification of animals and plants. Ray had reached two important conclusions: first, that the essential characteristics of each species of plant were determined by the seed from which it grew, and secondly, that a plant of one species never grew from the seed of another. This last conclusion gave him a criterion for distinguishing 'real kinds' in Nature. Two individual plants belonged to the same species *either* if the one had grown from a seed of the other, *or* if they shared a common ancestor—i.e. were related back genealogically to the same original source of seed. The boundaries between botanical species thus seemed clear and definite. Strictly speaking, no doubt, one could never establish *absolutely* that two plants which apparently belonged to different species had not had any common ancestor; but, with only six thousand years to consider, it was reasonable enough to think that this could be established beyond all practical doubt.

For practical purposes, therefore, most naturalists were happy to accept Ray's idea of a species. At a theoretical level, however, it encountered opposition. The standing criticism was directed, not against the *fixity* of the supposed species, but against the limitations which Ray was apparently placing on the infinite variety of organic forms. Though he himself rejoiced at 'the foecundity of the Creator's Power' displayed in the eighteen thousand known species of plants, perhaps one should not draw the line even there. Perhaps those eighteen thousand 'species' were themselves only taxonomic pigeon-holes, in a world whose richness and variety were beyond all numerical measure. As Bonnet put it:

> If there are no cleavages in nature, it is evident that our classifications are not hers. Those which we form are purely nominal, and we should regard them as means relative to our needs and to the limitations of our knowledge. Intelligences higher than ours perhaps recognize, between two individuals which we place in the same species, more varieties than we discover between two individuals of widely-separate genera. Thus these intelligences see in the scale of our world as many steps as there are individuals.

The accepted interpretation of the Great Chain of Being reinforced the belief that species are artificial: given the insensible gradations between adjacent steps in the Ladder of Nature, it seemed that organic nature must slip through the meshes of any merely human classification.

The only effective way to deal with these objections was to extend Ray's programme of classification throughout the biological kingdoms, and to see whether or not his idea of 'species' did apply to the world of Nature. This was what eighteenth-century zoologists were trying to do when they searched for 'natural' classifications as contrasted with 'artificial' ones. They hoped to mark off each kind of living creature by reference to a combination of characters which reflected its inner nature and reality. In a generation which still understood very little about the mechanism of reproduction, this was not easily accomplished, and throughout the eighteenth century the problem of species remained entangled with questions about spontaneous generation and preformation.

By the middle of the century, two men dominated the scene: Linnaeus in Sweden, and Buffon in France. These men represented the best-informed opinion of their time. Each set out in his own way to produce an encyclopaedic account of the world of Nature, and each left behind fascinating evidence of his intellectual development. Working from quite different positions—Linnaeus was committed from the outset to the idea of a systematic classification, while Buffon rejected it—they ended up in remarkably similar positions. To deal first with Buffon: the whole programme of the systematists was based, he declared, on an error in metaphysics—on 'a failure to understand nature's processes, which always take place by gradations'. At this stage (1749) he believed, like many of his contemporaries, that the continuity implicit in the Great Chain of Being was inconsistent with the reality of species—

> It is possible to descend by almost insensible degrees from the most perfect creature to the most formless matter. . . . These imperceptible shadings are the great work of nature; they are to be found not only in the sizes and the forms, but also in the movements, the generations and the successions of every species. . . . Nature, proceeding by unknown gradations, cannot wholly lend herself to these [taxonomic] divisions . . . there will be found a great number of intermediate species, and of objects belonging half in one class and half in another. Objects of this sort, to which it is impossible to assign a place, necessarily render vain the attempt at a universal system. . . . In general, the more one increases the number of one's divisions, in the case of the products of nature, the nearer one comes to the truth; since in reality individuals alone exist in nature.

In embarking on his monumental task of describing Nature, he was prepared to be quite candid about this limitation; but, in practice, he found this standpoint difficult to maintain and by the 1760s, though still rejecting all wider classifications as entirely arbitrary, he was prepared to accept species as a reality. Different species, he now declared, were 'as ancient and as permanent as Nature herself': by comparison, 'an individual . . . is nothing in the universe'. Beyond this he would not go. He was a good empiricist, and he regarded all attempts to impose a more elaborate system of classification (with genera, families, orders and so on) as prejudging questions which men were not yet in any position to answer—

> In general, the relationship between species is one of those profound mysteries of nature which man can only investigate by experiments which must be as prolonged as they will be difficult.

The relationship between the horse and the ass was, for him, a crucial test of the stability of species. Some people had argued that the ass originated as a degenerate form of the horse, and in 1753 Buffon set out to ridicule this view. Among a dozen objections, two stood out: first, the near-sterility of mules, which demonstrated 'the impossibility of uniting [the horse and the ass] to form a common species, or even an intermediate species capable of reproducing itself', and secondly, the ludicrous consequences which would follow, if one did not draw the line at this point. One could, in that case, equally well say that 'not only the ass and the horse, but even man, the ape, the quadrupeds and all animals might be regarded as making only one family'—and this view Buffon treated as plainly absurd.

By 1766 (when he published his essay on 'The Degeneration of Animals' in the fourteenth volume of the *Natural History*) Buffon was, however, prepared to contemplate the possibility that the two hundred existing species of quadruped sprang from some thirty-eight original stocks, the present-day forms having 'degenerated' (i.e. varied from the original forms) under the influence of changes in the environment. Once the red-hot Earth had cooled enough for life to appear upon it, the parent-stocks took shape spontaneously, their forms being determined by the 'moulds' or 'matrices' in which they were born. The present-day varieties are those which have survived the 'natural selection' described by Lucretius—

> At first the replacement of living Nature would be very incomplete, but in time all the creatures which were unable to reproduce would disappear; all imperfectly-organized bodies and all defective species would vanish; and there would remain, as there remain today, only the most powerful and complete forms, whether among plants or animals; and these new beings would be, in general, similar to the old ones because, the brute matter and the living matter remaining always the same, the same general plan of organization and the same varieties within particular forms would result.
>
> On this hypothesis one must suppose, however, that this new nature [of the creatures arising spontaneously at later times] would be shrunken, because the earth's heat is a power which affects the size of the moulds, and this heat being weaker today than it was at the beginning of Nature, the largest species could not arise or could not reach their present dimensions.

So long as the Earth's warmth was sufficient to generate them, Nature had continued to produce novel forms, though these were dwarfish as compared with the creatures of the first birth. But by now, large-scale spontaneous generation was no longer possible, since the requisite 'organic molecules' were not free to develop as they would have done in an unpopulated globe.

At this stage in Buffon's thought, a hint of evolutionary ideas had unquestionably entered his theories, and his middle-period convictions about the fixity of species were left far behind. Yet this overall vision of a cooling Earth spontaneously breeding a first generation of different organic stocks, from which later forms 'degenerated', was quite at variance with Darwin's. We can see in it at most a few fragmentary anticipations of Darwin, embedded in a setting borrowed from earlier natural philosophy; for Buffon's theory of the origin of species—like the more amateur efforts of Lord Monboddo and Erasmus Darwin (Charles Darwin's grandfather)—looked backward to such Greek thinkers as Empedokles and Lucretius, quite as much as forward to Darwin and Mendel.

One finds a similar development in Linnaeus' thought. Linnaeus was without question the greatest of the modern 'systematists'; and, by completing Ray's programme of classification, he put men in a position to go beyond it. Like Ray, he both sensed the multitudinous wonder of the natural world and felt the impulse to master it intellectually. His was the age of discovery. With every decade that passed, explorers were adding fresh species of plants and animals to the list of those already known, and the task of classification was becoming continually more urgent. In the absence of a satisfactory natural classification, Linnaeus made do with a comparatively artificial one, choosing as criteria of relationship (e.g.) the structures of the flowers and seed-pods, which form the reproductive organs of plants. And, as the discoveries accumulated, he worked continuously, in edition after edition of his *Systema Naturae*, to bring his own classification into accord with the full wealth of Nature. In all the early editions, right up to that of 1751, he insisted again and again that the existing species were products of the original Creation, and that time had not subsequently added to their number. Environmental differences might have affected the local stocks of any given species—so changing their stature, colour or fragrance—but these changes were, he believed, temporary and transient. Thus he set the seal of scientific authenticity on the theological belief in the stability of God's whole Creation, and by 1750 the belief in fixed species was widely believed to be an integral part of this broader doctrine.

Yet all the while Linnaeus' own experience was driving him towards a more moderate and flexible view. During the 1740s, he had been studying a quite new form of toadflax, which had apparently originated from the normal plant as a change of species, or (as he called it) a 'mutation', and which bred true to the new form. This 'peloric' form of toadflax—with its distinctive flower-heads—appeared in other parts of Europe also, and by now is regarded as resulting from a simple and frequent mutation, in our modern genetical sense of the word. Subsequently, Linnaeus studied many examples of apparently true-breeding hybrid plants, and by 1760 he allowed that

> It is impossible to doubt that there are new species produced by hybrid generation. . . . Many species of plants in the same genus in the beginning could not have been otherwise than one plant, and have arisen [from this] by hybrid generation.

By 1762, he seems to have thought that God had originally created the parents, not of each species, but of each genus or order, and that the different species had subsequently come into existence as variants of these original forms. Soon afterwards, he quietly dropped his fundamental assertion of the fixity of species—*nullae species novae*—from the last edition of his *Systema Naturae*, and by 1778 (the year of his death) he could no longer set limits to the possibilities of interbreeding and hybridization. Perhaps all species were 'the children of time'.

As with Buffon, one should neither ignore nor exaggerate the change in Linnaeus' viewpoint. Wherever he ended up, it was certainly not in a Darwinian position. Though he and Buffon disagreed about other things, they both took it for granted that the present population of living creatures had descended, with variations, from original parents at least as complex and highly-developed as themselves. These first ancestors might have been generated spontaneously out of organic molecules (as Buffon supposed), or specially created by the Almighty at the Beginning (as Linnaeus thought). Either way, the present population of living things had developed out of the ancestral stock by a mere process of variation. What neither of the great eighteenth-century naturalists imagined was that a primitive stock of extremely simple organisms could give rise progressively to a later population of more complex ones: this had always been the central stumbling-block to the theory of descent, and both Buffon and Linnaeus exchanged their earlier belief in the fixity of species only for a doctrine of organic diversification—of variations on an original theme.

The Temporal Sequence of Forms

If neither Buffon nor Linnaeus could face the possibility of organic evolution at all squarely, the reason lies in part in the scanty evidence they had about earlier forms of life. Both men were well informed about the *geographical distribution* of the existing species, and made excellent use of this knowledge in their arguments; yet—working before Werner and Hutton—they could have only a fragmentary acquaintance with fossil organisms, for this was all that yet existed. Both men had been born in

1707: Linnaeus died in 1778, Buffon ten years later. But even at eighty, Buffon was in no position to reconstruct the temporal succession of organic forms from the sequence of fossils in the Earth's crust, or to recognize the progressive increase in complexity of organisms through geological time. At the time of his death, the burst of field-work initiated by Werner was only beginning, and it was twenty years and more before the work of Lamarck, Cuvier and William Smith created a genuine science of palaeontology, in which the fossil remains of extinct organisms were pieced together, compared and set in a chronological order.

In the long run, the palaeontological evidence played an essential part in establishing the theory of evolution, but its first effects were paradoxical. Lamarck's evolutionary theory of organic change was part of an overall philosophy of Nature: he used his palaeonotological studies merely to illustrate and confirm his basic ideas. Cuvier, preferring to keep closer to his own experience, argued directly from the palaeontological evidence, and ended by creating a prejudice against the idea of evolution which was still active among French zoologists thirty years after his death.

Both men were strongly influenced by Buffon, and Lamarck had actually worked with him. Lamarck saw, more clearly than Buffon had done, the link between the problem of the origin of species and the problem of geological change. The fundamental question in each case, he argued, was how the raw materials of the world had acquired its present organization. He was sceptical about Buffon's suggestion that the first generation of animals was formed directly from 'organic molecules' in terrestrial 'matrices'—an idea which found no support in the fossil records. One should reject any absolute distinction between organic and inorganic matter, and look for the common principles responsible for organization in both realms. Granted that the Earth had begun as a formless, fiery mass, the evolutionary question was inescapable: by what sequence of natural processes had the original formless matter of the world been brought to its present level of physical, chemical and biological organization?

So far, Lamarck's argument reads like twentieth-century commonsense. Unfortunately, he immediately became involved in a dispute about the principles of chemistry in which he set himself up, gallantly but foolishly, against the most experienced chemists of his time. (On this dispute, see *Architecture of Matter*, chapter 10.) The battleground was ill-chosen, and the chief victim was Lamarck's own reputation. He suffered from the effects of this self-inflicted wound for the rest of his life, and even after his death his support for the doctrine of evolution was an obstacle to its acceptance in France. Right up to 1860, everything played into the hands of his sound and politic colleague, Georges Cuvier. The situation contained elements of tragedy. For, just when Lamarck's general philosophy was most thoroughly discredited, he at last found a field of study in which it began to yield positive results. To begin with, he had been closely concerned to understand how a universal process of evolution gave matter its chemical properties, and so had paid little attention to evidence of geological change; but in the 1790s, when he was nearly fifty, he was appointed to the Invertebrate Division of the Paris Museum. At last he had a positive subject in which the speculative power of his imagination could be put to effective use, and during the next twenty-five years he used his first-hand knowledge of the invertebrate fossils from the chalk around Paris to display the 'gradual perfecting of the organization' of living things, through periods of duration 'which overwhelm our thought'.

In appraising Lamarck's contribution to the evolutionary debate, one must distinguish questions of three kinds: those of historical fact, those of broad interpretation and those of mechanism. About the fundamental facts revealed by palaeontology, Lamarck's position is by now unanswerable. The older the rocks, the simpler the forms of life they contain: traces of the higher animals are confined to geological strata of comparatively recent date. So the doctrine of organic *progression*, for which Herder had argued on philosophical grounds, won support from palaeontology. (On this point, there was no disagreement between Lamarck and Cuvier, who was an expert on fossil vertebrates.) But, when it came to interpreting this progression, Lamarck was a Cassandra. The temporal succession of organic forms—gradually increasing in complexity from one geological epoch to the next—was for him clear evidence that the later animals were related to the earlier by actual descent. Granted only a sufficient period of time, the idea of an evolutionary genealogy had an almost mathematical plausibility. There was no more need of 'special creations' or ubiquitous 'organic molecules', and the mysteries which had disfigured the theories of Buffon and Linnaeus could be swept away. The whole problem of the relationship between species which to Buffon had seemed almost too deep to fathom, could be restated in two simple questions. First: How the creative action of Nature originally gave organization to the basic raw material; and secondly. What agencies were responsible for the progression from one level in the scale of organization to the next higher and more complex level. Confident that these were the fundamental questions, Lamarck published in his *Zoological Philosophy* (1809) the first evolutionary family-tree, showing the branched series by which the complex organisms of the present day were related back to earlier, simpler forms of life, and so to the hypothetical point of the 'first beginnings of organization'.

Lamarck's two questions were not easily answered. In the 1820s one could scarcely say anything serious about the origin of life: lacking even the rudiments of biochemistry, Lamarck attributed it to 'heat and humidity'. Over the second, and even more crucial question of mechanism, Lamarck's impatience betrayed him again. He seems to have thought that, once the hypothesis of descent was admitted, a dozen possible mechanisms might account for the appearance and survival of new and more complex forms: he did not apparently realize that the very acceptability of the theory of descent must depend on establishing the *actual* mechanisms involved. Speculative as ever, he handled the question in the spirit of Lucretius and filled his books with feebly-supported explanations of more and more mechanisms, rather than critical proofs of any one. Sometimes, he pointed to external facts as chiefly responsible:

> *Species* have only a limited or temporary constancy in their characters, and . . . there is no species which is absolutely constant. Doubtless they will subsist unchanged in the places in which they inhabit so long as the circumstances which affect them do not change, and do not force them to change their habits of life.

Sometimes, the development of new limbs or organs was the more or less unconscious response of animals coping with the new modes of life in a new environment. Sometimes, he gave the credit for organic progression to the 'universal force' responsible for all organization in Nature. Just occasionally, he mentioned competition between species as helping to diversify animal structure and varieties. And so he went on, multiplying plausible mechanisms of organic variation, without making any one of them really convincing.

The trouble was that few of his contemporaries were prepared, even for the purposes of argument, to grant his fundamental hypothesis of descent. Generally speaking, they would admit no more scope for variation than Buffon and Linnaeus had envisaged and some, such as Cuvier, recognized the existence of varieties only within a given species. Even those who accepted the mutability of species still found it easier to accept a theory based on organic degeneration than one in which the higher animals were progressively formed from the lower. It was no use Lamarck making his theory attractive by talking in general terms about use and disuse, adaptive variation and the influence of the environment—still less, by invoking the 'universal creative principle'. The more he multiplied his explanations, the more suspicious his critics became. What offended them was not his assumption of the inheritability of acquired characteristics—the standard objection to Lamarckism today—but the intellectual gap that still remained between his evidence and the theory he used to support it. They would be convinced that organic progression *could* be accepted as evidence of evolutionary descent, only when they were convinced that it *must* be; and it was this kind of certainty that Darwin's work was later to provide. Lamarck himself started with a prejudice in favour of evolution: for most of his contemporaries the burden of proof was reversed, and the vagueness of Lamarck's arguments merely confirmed their opposition.

This was particularly true of his successful rival, Cuvier. The contrast between the careers of the two men contains, indeed, a good deal of dramatic irony. Jean Baptiste Pierre Antoine de Monet, Chevalier de Lamarck, started with all the obvious advantages of birth and education, and was launched on his career by the great Buffon himself; yet for the rest of his life he seemed to fumble his chances, and died in his mid-eighties blind, despised by his colleagues, and with his vision of nature discredited. His premature support for uniformitarian geology, with its extended time-scale, earned him abroad the title of 'the French atheist'; and his eager support for the French Revolution only confirmed this judgement. By contrast, Georges Cuvier, though an 'outsider', scarcely put a foot wrong throughout his whole lifetime. He was a Protestant of Swiss ancestry, and was educated at Stuttgart. Yet from the moment he arrived in Paris in 1795, he entered on a career of unchecked success, as one of the 'new men' of the post-revolutionary era. In addition to his academic success and his polished literary style, his administrative talents were outstanding and he was appointed official spokesman in the French Parliament, first by Napoleon, and later by Louis XVIII. As secretary of the *Académie Française* he dominated the intellectual life of France, and his geological theory of catastrophes was seized on internationally as a way of 'rescuing geology from the atheists'.

Cuvier's field of study—the fossil vertebrates—was far more spectacular than Lamarck's humbler field. Mammoths, mastodons, extinct forms of rhinoceros, hippopotamus, deer and crocodile: in Cuvier's hands, the vanished giants of a former epoch took shape again, and the whole learned world marvelled at the richness of this unsuspected fauna. By the end of his life, he was so familiar with the skeletal structure of the vertebrates that, given a single fossilized bone of an unknown species, he could (he boasted) argue his way to the whole animal (Plate 6). He died in 1833, four years after Lamarck, leaving behind as his life-work twelve large volumes of the fossil bones of quadrupeds. The shadow of his ideas, notably his outright rejection of evolutionary descent, lay over French science for many years.

In retrospect, Cuvier's scientific achievements inevitably look like successful delaying-action: we know now that the attractiveness of his ideas was specious and his crucial evidence misleading. But in the context of his time, the strong points of his position were far from negligible, and there were certain aspects of the problem of species which he appreciated better than Lamarck. For Lamarck had concentrated too exclusively on the highest forms of life in any given era—the 'spearpoint of evolution'—

and tended to ignore the fate of those simpler forms which had come into existence in earlier epochs. Presumably, they had survived in some form or other, adapting themselves where necessary to the changing environment. Nature being for him the universal Creator, the possibility of their being entirely wiped out ran against the grain of his thought:

> One may not conclude that any species has really been lost or annihilated. It is doubtless possible that among the largest animals there have been some species destroyed as a result of the multiplication of man in the places which they inhabit. But this conjecture cannot be founded solely upon the consideration of fossils; one cannot pronounce in this matter until every habitable part of the globe shall be perfectly known.

Cuvier, with his superior knowledge of the higher animals, was quite prepared to 'pronounce on this conjecture': vertebrate palaeontology showed clear evidence that whole species, and even genera, had repeatedly become extinct. So it was not enough to explain how the lowest forms of life had originated, and by what mechanisms new and more complex organisms were produced in each epoch: one must also explain how living species *of every kind* are so perfectly adapted to their present environments, and why similar species from earlier epochs have disappeared from the Earth.

This was a good question, for a consideration of the relation between adaptation and extinction was one of the crucial points from which Darwin was to start. Cuvier's own answer, however, was of a piece with his geological theories. The discontinuities between strata which had driven Cuvier the geologist to the idea of 'catastrophes' were apparent in the fossil record equally, creating a problem which Cuvier the palaeontologist dealt with in exactly the same way. If the fossil species from geological strata of different ages were so different from each other, then the actual animal populations inhabiting the Earth must have changed strikingly from epoch to epoch. Yet there was no more evidence for a smooth transition from one zoological population to the next than there was for gradual geological transitions from stratum to stratum: on the contrary, all field studies pointed to a series of violent contrasts between long eras of constant conditions and abrupt, perhaps even instantaneous, transformations. Every epoch had apparently seen a different pattern of continents and oceans, a different structure of mountains and plains: every epoch, too, had been characterized by its own distinct fauna, and the catastrophe which re-shaped the Earth at the end of each epoch also extinguished, largely if not completely, the animal species of the earlier age.

> Life on this earth has accordingly frequently been afflicted by these terrible occurrences. Numberless living things have been the victims of these catastrophes: those living on dry land having been swallowed up by floods, others whose home was in the waters being left high and dry when the sea-beds were suddenly lifted up. Whole races have been wiped out for ever, leaving no record in the world but those few relics which even the naturalists can barely recognize.
>
> Such are the conclusions to which we are necessarily led by objects which we encounter at every step, which we can verify whenever we please in nearly all countries. These great and terrible events have everywhere left an imprint clearly visible to anyone who knows how to read its story in their monuments [i.e. the rocks].

Cuvier went further. Evidence for the continuity of living species was not merely lacking: it was not to be expected. He had adopted precisely those of Buffon's ideas which Lamarck had abandoned—notably, a belief that hybrids were normally sterile, and that any extremes of variation appearing in individuals would tend to be eliminated in subsequent generations. First-hand experience teaches us, he argued, that a stock of animals and plants remains vigorous only if cross-breeding is restricted, and presumably this has always been so. Accordingly, the novel fauna of each new geological epoch could not be descended from the animals and plants of the former epochs, but must represent fresh creations, ready adapted to the novel environment. That was true, most of all, in the case of the human race: Man was freshly created only a few thousand years ago, surviving the last catastrophic Flood thanks to the intervention of divine Providence.

Yet Cuvier's theoretical twist created a new difficulty. According to Lamarck's theory of descent, mammoths and elephants were genealogically related, and this made their similarities intelligible enough: Lamarck's difficulty was to produce a convincing explanation of organic *change*. Cuvier's problem was the reverse: if there were renewed organic creations after every catastrophe, that accounted for the difference between the animals of different epochs, at the cost of leaving their *resemblances* mysterious. Having rejected genealogy, Cuvier had to look for the required explanation elsewhere, and for this purpose he revived an earlier—and now forgotten—view of Nature. In every age, the fossil record revealed species belonging to four recognizable 'branches': the vertebrates, the molluscs, articulated creatures such as the insects, and an assortment of radially-symmetrical creatures which he called 'radiata'. This fourfold division in the fundamental structures of animals was, Cuvier taught, an essential feature of the 'ideal specification' according to which Nature fashioned all her creations. Early in each geological epoch, Nature 'called into existence' creatures adapted for life in the new environments and her creative activities were always guided in conformity with the same four basic plans. However much the geological kaleidoscope was shaken up, these four fundamental patterns of the Animal Creation were preserved from one epoch to the next.

This aspect of Cuvier's theory had great theological attractions for the public of his time. If the geological catastrophes were supernatural interruptions in the Order of Nature, so too were the subsequent re-creations of organic life; and Cuvier now added the four basic patterns of animal life to the previous evidence about the rational specification realized in the Divine Creation. Yet, scientifically speaking, Cuvier was in this way evading a question to which men were bound later to return. In their different ways, Buffon and Lamarck had both tried to explain how the first generation of living things originated: Lamarck accepting spontaneous generation in the case of the lowest forms of life, Buffon supposing that higher animals were originally formed direct from 'organic molecules'. Cuvier, however, left the process by which each new fauna and flora came into existence entirely vague. This problem was exceedingly difficult on any account—as William Whewell pointed out in his review of Lyell's *Principles of Geology*:

> To give even a theoretical consistency to his system, it would be requisite that Mr. Lyell should supply us with some mode by which we may pass from a world filled with one kind of animal forms, to another, in which they are equally abundant, without perhaps one species in common. . . . From the plesiosaurs and pterodactyls of the age of the Lias, to the creatures which mark the Oolites or the iron-sand. He must show us how we may proceed from these, to the forms of . . . the paleotherian and mastodontean periods. To frame even a hypothesis which will, with any plausibility, supply this defect in his speculations, is a harder task than that which Mr. Lyell has now executed.

Still, once Lyell had demonstrated that Cuvier's geological discontinuities could be explained quite easily in uniformitarian terms, the corresponding discontinuities in Cuvier's palaeontology became a standing challenge to which, in time, uniformitarian zoologists were certain to turn their attentions.

The Problem of Inheritance

Up to this point we have concentrated on the two main strands of evidence which were available around 1800 to zoologists concerned with the problem of species: namely, the geographical distribution of existing species in the Old World and the New, and the temporal succession of living forms revealed by the fossil record. But these represent only part of the material which must be accounted for in any explanation of the origin of species. Certainly, the distribution of species in space and time is a crucial element in the problem: in a sense, it is the primary phenomenon. How the species which now exist in different parts of the world have come to resemble—and differ from—one another as they do; and what connection, if any, they have with those other extinct species which preceded them: taken together, these two questions pose the central *historical* problem of species. But in addition to these primary historical questions, the origin of species has always involved certain other consequential problems: problems of inheritance, embryological development and general physiology. It is interesting to find, therefore, that as early as 1750 some scientists at least appreciated these consequential issues much better than they did the central historical problem.

Two such questions which are closely connected with the wider problem had been thoroughly debated: the problem of spontaneous generation, and the nature of embryological inheritance. The theories of Buffon and Cuvier about the first beginnings of species depended ultimately on a belief in the spontaneous generation of the higher animals. (This took place either when the Earth had cooled down sufficiently, or repeatedly after each catastrophe.) Yet it had been established well before 1700, by Redi and Swammerdam, that all the reputed kinds of spontaneous generation—of mice in dirty shirts, maggots in putrid meat and so on—were spurious; so that many physiologists took it as established that, at the present time, living creatures were born only from others of their own kind. Only at the lowest end of the scale did doubt remain: the possibility that the simplest forms of protozoa might be generated directly out of non-living raw materials could not be ruled out. (See chapter 14 of *The Architecture of Matter*.) It was the publication of Virchow's generalized cell-theory, almost simultaneously with Darwin's *Origin of Species*, which eliminated spontaneous generation, even at the cellular level, as a continuing element in the creation of living things.

It would seem at first sight that a strong point of Lamarck's system was that it, too, dispensed with spontaneous generation except for the most rudimentary forms of life. Yet this point was not at first appreciated. Those who believed in 'biogenesis'—the doctrine that living things are born only from similar living things—saw this as confirming the fixity of species. By the late eighteenth century, the power of organisms to reproduce their own kind was widely explained in terms of certain preordained, preformed seeds or germs, each generation containing implicitly the elements of all its successors. This conviction, that all forms of life were essentially determined by the pre-established character of their seeds, told strongly against ideas of organic transformation. Indeed, the final acceptance of evolution was to be dependent on a better understanding of embryology, and this grew from the work of von Baer from 1827 on; while the modern synthesis of genetics and evolution-theory was a long way further off. Yet many of the lines of thought which eventually led to this synthesis had already been argued out in the 1740s and 50s, by Pierre Louis Moreau de Maupertuis, the brilliant and versatile scientist who for more than ten years was in charge of Frederick the Great's Academy of Sciences at Berlin, and whose observations on nebulae were the starting-point of Kant's cosmology (see Chapter 6, above).

Maupertuis was unfortunate in having his reputation torn to shreds at a critical moment by the satirical pen of Voltaire, who had formerly been his great admirer. In 1752, after a bitter quarrel about the direction of the Berlin Academy, Voltaire seized on two open letters in which Maupertuis had proposed a series of original scientific investigations, and used these in his *Diatribe du Docteur Akakia* to launch an attack of the most sarcastic savagery, from which Maupertuis' reputation never fully recovered. Maupertuis had, nevertheless, grasped many of the essential genetical and embryological points about the mechanism by which the identity of species is first maintained, and then gradually altered in the course of evolution. Up to that time, no mechanism had been envisaged by which both parents might make equal contributions to the character of their offspring. Maupertuis himself refused to be stampeded by the hypothetical 'animalcules' seen with the newly-invented microscope into denying what every animal-breeder knew very well—the facts of bi-parental heredity. Instead, he took every opportunity to investigate the pattern of inheritance directly, and explained his results by a genetical mechanism strikingly like the modern 'chromosome' theory. (In a final query, he even anticipated that such a mechanism might provide a means for the production of new species.)

Maupertuis' studies of inheritance were of two main kinds. The first involved a systematic investigation of certain inheritable human characteristics. The arrival in Paris of a young albino negro from Africa had been a nine days' wonder in France. Instead of merely applauding, Maupertuis was intrigued. According to reports, this albinism was not uncommon in Senegal, where it tended to run in certain families. So, Maupertuis decided, it must be an inherited trait in negroes just as in birds; and, if one could trace the mode of inheritance of such a striking peculiarity, this should tell us something significant about the mechanism of heredity. Arriving in Berlin, he set enquiries in train, and he was soon introduced to a surgeon, Jacob Ruhe, who had been born with an extra digit on each hand and foot. Though his father had been perfectly normal, Ruhe shared this peculiarity with his mother and his maternal grandmother, while three of his seven brothers and sisters were also affected: as for his children, two out of six had extra digits, although his wife was normal and had no corresponding family history of abnormality. As Maupertuis saw, this example had two important consequences. Such 'polydactyly' was both striking and rare, occurring on his estimate in one in twenty thousand people at the most. Chance alone could scarcely explain its reappearance in three successive generations of the Ruhe family, for the odds against this happening even once in each generation were some eight million million to one—'a number so great that the certainty of the best demonstrated theories of physics does not approach these probabilities'. Furthermore, one could hardly have a better demonstration of biparental inheritance, for the abnormality had been transmitted first through the female line and subsequently through the male.

From this, Maupertuis went on to actual breeding experiments, using animals of all kinds. Though he could never deal with the numbers required for a full statistical analysis, such as Mendel was able to do, he formed some remarkably clear ideas about the pattern of inheritance, and about the underlying mechanism which might account for it. In the process of sexual reproduction, he concluded, both the father and the mother contributed a set of 'particles'. Corresponding particles from each parent had a special 'affinity' for one another, comparable to the 'elective attractions' of eighteenth-century chemistry: and, having united, each pair then played a part in determining the development of the offspring. In any pair, either the paternal or the maternal particles might be dominant, so that a peculiarity transmitted by a single parent might reappear only after several generations. The particles were generally stable, but occasionally abnormal ones might appear, whose influence could be made dominant by artificial selection. In this way, men were every year creating

> races of dogs, pigeons, canaries, which did not at all exist in Nature before. These were to begin with only fortuitous individuals; art and repeated generation have made species of them.

How these fortuitous changes came about, Maupertuis was not sure. Perhaps the environment had an effect—

> Although I suppose here that the basis of all these variations is to be found in the seminal fluids themselves, I do not exclude the influence that climate and foods might have. It seems that the heat of the torrid zone is more likely to foment the particles which render the skin black, than those which render it white. And I do not know to

> what point this influence of climate or of foods might extend, after long centuries of time.

Nevertheless, he saw clearly the direct connection between the problem of genetic inheritance and the problem of species, and at the end of his *System of Nature* he threw out a question which was to be prophetic:

> Could one not explain by that means [the fortuitous appearance of mutant 'particles'] how from two individuals alone the multiplication of the most dissimilar species could have followed? They could have owed their first origination only to certain fortuitous productions, in which the elementary particles fail to retain the order they possessed in the father and mother animals; each degree of error would have produced a new species; and by reason of repeated deviations would have arrived at the infinite diversity of animals that we see today; which will perhaps still increase with time, but to which perhaps the passage of centuries will bring only imperceptible increases.

Maupertuis' genetical ideas were discussed seriously by Buffon and Bonnet, but soon afterwards forgotten, being rediscovered only in the twentieth century. One may be tempted to blame Voltaire for this eclipse, yet perhaps he was not entirely responsible. For Maupertuis was one of those men who are not content to remain within the boundaries of certainty, or to push forward the frontiers of science one small step at a time. He spent his whole career staking intellectual claims far beyond the frontiers of his own time, and sketching the form of problems which few of his contemporaries were ready to tackle. Faced with such a man, one is always tempted to ask, 'What might he not have achieved if . . . ?'—if modern geology had developed seventy-five years earlier, for example, might not Maupertuis have anticipated both Darwin and Mendel, and so placed evolution-theory on a genetical basis from the very beginning? But we learn more from a study of what actually happened than from speculating about what might have been. What is important about Maupertuis is not just the penetration of his analysis, or the far-sightedness of his theories, but the very fact that his contemporaries *did not appreciate them*. Biologists in the mid-eighteenth century were intellectually unprepared for the implications of evolution; for physiology was still rudimentary, biochemistry non-existent, embryologists were bogged down in disputes about preformation, inheritance-studies were a complete novelty, and—above all—the obstacles presented by the pre-geological time-scale were overwhelming. Maupertuis was thus arguing in a vacuum, and little of what he said made intelligible sense to his colleagues.

The Final Impasse

As we have already seen, the last and most obstinate of these difficulties was removed in 1830, by Charles Lyell's uniformitarian geology. Yet nearly thirty years were to run before the next step was taken, by which the zoological account of the development of organic species became as genuinely historical as the geological account of the development of the Earth's crust. During these thirty years, scientists were divided into rival parties, none of which appears, by our standards, to have been entirely in the right; yet no one at the time saw how to take the next step forward without making damaging concessions to the opposite camp. The resulting deadlock was broken only by Charles Darwin.

How did the theoretical division run? The key figure, once again, is Charles Lyell. Darwin owed more to Lyell's *Principles of Geology* than he did to any other single book—even Malthus' *Essay on Population*; he was happy and proud to serve as Lyell's scientific assistant, and to provide further examples from his own geological experience to confirm the soundness of Lyell's principles; and to Darwin, as to us, the evolutionary explanation of the origin of species appeared to be a natural continuation of Lyell's own 'uniformitarian' argument in geology. But that was not how Lyell himself saw it. In his eyes, the uniformitarian axiom implied more than it was to do later, either to Darwin or to us. Speaking in old age about his own first reaction to the *Origin of Species*, he still regarded his eventual conversion to Darwin's theory, not as the fulfilment of his earlier vision, but as its abandonment: 'It cost me a struggle to renounce my old creed.'

Like everyone involved during the 1830s and 40s, Lyell had one eye fixed on the geological evidence, while keeping the other focussed on what his opponents were doing. The resulting double-vision led him into difficulties from which he could not escape unaided. For he felt compelled to reject, as irrelevant to geology, both the idea of organic progression, and the theory that the Earth was slowly cooling. The trouble was partly technical, partly ideological. His own fundamental conviction was 'that all former changes of the organic and inorganic creation are referrible to one uninterrupted succession of physical events, governed by the laws of Nature now in operation'. This he interpreted as implying a uniformity both of causes and effects. He supposed, first, that the same physical *causes* have shaped the Earth and its inhabitants at all times, with a force barely exceeding the limits still to be observed today; and, going further, he assumed in addition—failing unanswerable evidence to the contrary—that the *effects* of these agencies in all epochs were as constant and uniform as the agencies themselves. On this count alone, he argued, both organic progression and Buffon's refrigeration-theory had to be rejected. To deal with the latter theory first: there was, Lyell conceded, a certain plausibility about Buffon's argument, which the later calculations of Fourier had done something to increase. Yet no positive evidence could be found that this cooling, even if it was authentic, had left any detectable mark on geological phenomena—still less that it was, as some believed, 'the principal cause of alterations of climate':

> On the contrary, La Place has shown, by reference to astronomical observations made in the time of Hipparchus, that in the last two thousand years there has been no sensible contraction of the globe by cooling; for had this been the case, even to an extremely small amount, the day would have been shortened, whereas its length has certainly not diminished during that period by 1/300th of a second.

Nor was Fourier's argument that the Earth must be cooling, since it was continually losing heat by radiation, entirely conclusive:

> This argument may appear plausible, until we reflect how ignorant we are of the sources of volcanic heat, or indeed of the nature of light and heat in general. It is doubtless true, that light and heat are con-continually emanating from the earth; but, in the same manner, it may be said that they escape without intermission from the sun, and we know not whether there be any compensating causes which again restore them to that luminary.—'It is a mystery,' says Herschel, speaking of the sun, 'to conceive how so enormous a conflagration (if such it be) can be kept up. Every discovery in chemical science here leaves us completely at a loss, or rather seems to remove farther the prospect of probable explanation.'

With the example of the Sun before us, we could not lightly assume that all the sources of terrestrial heat were already known. On the contrary, all Buffon's and Fourier's 'curious experiments on the cooling of incandescent bodies' might yet prove irrelevant to the theory of the Earth, if the heat which it radiated away was continually made good by subterranean processes hitherto unknown. Here, of course, Lyell's caution was well founded: the contribution of radioactive heating to mountain-building, like the thermonuclear source of the Sun's heat, was not identified until the twentieth century, and meanwhile the time-scale required for geological development and organic evolution remained under a question-mark.

But Lyell's objections to the refrigeration theory were not simply technical. For Buffon's theory had been given a twist by Cuvier and the

Catastrophists which made it obnoxious to a good Uniformitarian like Lyell. In short: the near-collision by which—on Buffon's account—a passing comet had struck the matter of the planets off from the Sun was just the sort of violent and hypothetical cataclysm that suited Cuvier's own purposes. Whereas in 1750 the idea of this initial catastrophe had offended theologians as being excessively mechanical, by the 1820s it had a fresh appeal to devout geologists. For it both represented a dramatic demonstration of the divine Power, and also had a providential flavour which they found very welcome: though the slow cooling of the Earth's mass might in itself be a purely physical process, it was, all the same, nicely contrived to create an environment in which living things—and eventually Man—could be given a home.

Lyell's objections to the idea of organic progression were also stimulated, in part, by its apparently providential implications. When, in the 1780s, Herder had argued that more and more complex organisms had appeared on the Earth in succession, he presented this progression as an essential element in 'Nature's plan', leading up to the appearance of Man, who 'was to possess the Earth and be the Lord of Creation'. Developed further by Schelling and Oken, Herder's system of thought gave rise to the biological theories known as 'Nature-philosophy': according to these, the present state of Nature was the realization of certain 'ideal plans' which served as blueprints for all Nature's creative operations. The Nature-philosophy of Schelling and Oken had an affinity with Cuvier's ideas, and in the person of Louis Agassiz, who completed Cuvier's half-finished work on the fossil fishes, the two traditions were united. Before beginning his collaboration with Cuvier, Agassiz had been a student of Oken's at Munich, and for the next thirty years he was the leading spokesman both for Catastrophism and for Organic Progression. Both doctrines, in his view, provided evidence of God's plan for the world.

Reacting against this alliance of progression, catastrophism and providentialism, Lyell was highly suspicious of 'the theory of the successive development of animal and vegetable life, and their progressive advancement to a more perfect state'. On the face of it, this theory was directly opposed to his own fundamental belief in 'one uninterrupted succession of physical events, governed by the laws of Nature now in operation'; and he scrutinized the geological evidence for progression with savage care. Sir Humphrey Davy, for instance, had argued that

> In those strata which are deepest, and which must, consequently, be supposed to be the earliest deposited, forms even of vegetable life are rare; shells and vegetable remains are found in the next order; the bones of fishes and oviparous reptiles exist in the following class; the remains of birds, with those of the same genera mentioned before, in

> the next order; those of quadrupeds of extinct species in a still more recent class; and it is only in the loose and slightly consolidated strata of gravel and sand, and which are usually called diluvian formations, that the remains of animals such as now people the globe are found, with others belonging to extinct species. . . . There seems, as it were, a gradual approach to the present system of things, and a succession of destructions and creations preparatory to the existence of man.

Davy's evidence was not good enough to convince Lyell. The doctrine of progression, he replied, 'though very generally received, has but a very slender foundation in fact'. The same caution which had led him to deny the 'catastrophist' significance of Cuvier's geological discontinuities now encouraged him to suspend judgement once again. For the mere fact that the fossils *surviving from* earlier and earlier geological epochs belong predominantly to simpler and simpler types, was not by itself evidence that the animals and plants *existing in* those epochs belonged exclusively to those same types. The chances of any particular organism surviving into later epochs as a fossil depended on a dozen factors:

> Almost all the animals which occur in subaqueous [geological] deposits are such as frequent marshes, rivers, or the borders of lakes, as the rhinoceros, tapir, hippopotamus, ox, deer, pig, and others. Species which live in trees are extremely rare in a fossil state; and we have no data as yet for determining how great a number of the one kind we ought to find, before we have a right to expect a single individual of the other.

If all the evidence were surveyed critically, it was impossible to establish conclusively any general tendency or direction of change:

> In regard to plants, if we neglect the obscure and ambiguous impressions found in some of the oldest fossiliferous rocks, which can lead to no safe conclusions, we may consider those which characterize the great carboniferous group as the first deserving particular attention. They are by no means confined to the simplest forms of vegetation, . . . but, on the contrary, belong to all the leading divisions of the vegetable kingdom. . . .
>
> If we then examine the animal remains of the oldest formations, we find bones and skeletons of fish in the old red sandstone, and even in some transition limestone below it; in other words, we have already vertebrated animals in the most ancient strata respecting the fossils of which we can be said to possess any accurate information. . . .
>
> [As for the quadrupeds,] we cannot detect any signs of a progres-

> sive development of organization,—any indication that the Eocene fauna was less perfect than the Miocene, or the Miocene, than . . . the newer Pliocene.

True, Lyell conceded—in the earlier editions of the *Principles*—Man had appeared on the geological scene only very recently. Yet even that fact did not subvert his Uniformitarian principles, or lend weight to the doctrine of progression:

> The superiority of man depends not on those faculties and attributes which he shares in common with the inferior animals, but on his reason, by which he is distinguished from them. When it is said that the human race is of far higher dignity than were any pre-existing beings on earth, it is the intellectual and moral attributes only of our race, not the animal, which are considered; and it is by no means clear, that the organization of man is such as would confer a decided pre-eminence upon him, if, in place of his reasoning powers, he was merely provided with such instincts as are possessed by the lower animals.
>
> If this be admitted, it would by no means follow, even if there had been sufficient geological evidence in favour of the theory of progressive development, that the creation of man was the last link in the same chain. For the sudden passage from an irrational to a rational animal is a phenomenon of a more distinct kind than the passage from more simple to the more perfect forms of animal organization and instinct. To pretend that such a step, or rather leap, can be part of a regular series of changes in the animal world, is to strain analogy beyond all reasonable bounds.

These quotations express Lyell's point of view during the early 1830s. In the next twenty years, geologists from both camps contributed to the progress of their common science. Louis Agassiz, the Catastrophist, established the occurrence of earlier Ice Ages, during which prolonged frost had shattered the surface rocks, and vast boulders had been transported by glaciers far from their point of origin. Charles Darwin, as a Uniformitarian, was able to strengthen the weak points in Lyell's theory of mountain-building, by reporting on two earthquakes that he witnessed for himself in Chile, by which the foothills of the Andes were lifted through some dozens of feet in a single movement. (Such a change was decidedly more violent than anything quoted in the first edition of Lyell's *Principles*.) So, between 1830 and 1850, the contributions from both schools of geologists interlocked and supplemented one another. The 'uniform agencies' of which Lyell had evidence gradually increased in force, until they became almost 'catastrophic' in their possibilities, and

Agassiz's 'catastrophes' meanwhile became so much smaller and more regular that they developed a 'uniformity' of their own. Over points of fundamental doctrine, however, the differences between the schools remained as sharp as ever. As late as 1851, Lyell was still denying the evidence for organic progression. 'There is no manifest elevation in the grade of organization, implying a progressive improvement in the flora [or fauna] which succeeded each other from the Eocene to our own epoch.' Nothing which had happened since the first appearance of the *Principles* had led him to change his view. So far as the problems of origins was concerned, geology as he saw it was still where James Hutton had left it:

> In the first publication of the Huttonian theory, it was declared that we can neither see the beginning nor the end of that vast series of phenomena which it is our business as geologists to investigate. After sixty years of renewed enquiry . . . the same conclusion seems to me to hold true.

So long as opinion was divided along these lines, the idea of organic descent could win no serious foothold in science. For the theory of descent combined two apparently incompatible beliefs: the denial of supernatural causes (which was a Uniformitarian axiom) and the doctrine of organic progression (which was a Catastrophist speciality). Right up to 1859, the majority of scientists in both Europe and America could accept *either* the uniformity of natural agencies *or* the idea of progressive succession, *but not both*. Meanwhile—Darwin's private thoughts apart—the cause of evolution was pleaded by a small minority, who were commonly dismissed as cranks or crackpots. Matters were not improved when, in 1844, an amateur polymath called Robert Chambers (the editor of *Chambers' Encyclopaedia* and a dozen other popular handbooks) published anonymously a defence of evolution under the title of *Vestiges of the Natural History of Creation*. For this book, which shared many of the fundamental weaknesses of Lamarck's *Zoological Philosophy*, was widely read and universally condemned, and thus strengthened the disrepute in which the theory of descent was already held. So far as most serious scientists were concerned, the idea of evolution had become, by 1850, a non-starter.

The task facing Darwin was, accordingly, one of selection and reconciliation. In point of method, he was convinced by Lyell's insistence on the uniformity of natural causes: yet, at the same time, he found the evidence for organic progression more convincing than Lyell had ever done, and he recognized very early that a genealogical relationship between the species of different epochs would provide the natural connection which his uniformitarian principles demanded. The outstanding problem was that which Lamarck had failed to solve—the problem of *mechanism*: to explain how, within the limits of geological time, the minute variations known to occur between the individuals of different species could have become dominant, and accumulated, to such an extent that they repeatedly gave rise to new and distinct species.

The intellectual deadlock in which the scientist of the 1830s and 40s were joined originated, accordingly, only partly in the evidence and technical arguments of geology. As always, the scientific evidence had to be interpreted, and the rival interpretations were frequently affected by far more intangible considerations. Behind the opposed conclusions of Lyell and Agassiz, one can detect the broader, philosophical stances in which they faced the evidence. Something of Newton and Hutton survived in Lyell's ways of thought: he would always opt for a stable equilibrium of natural forces, preserving on Earth the same uniform Order of Nature that characterized the planetary system. By contrast, Cuvier and Agassiz always hoped to find a place in the history of Nature for supernatural interventions: on their view, divine Providence had established the present order, not all at once, but by a progressive succession of creative acts. Finally, Lamarck has been described as 'an eighteenth-century deist': his genealogical succession of organic creatures was the product, over a vast period of time, of two opposed natural forces—one destructive, the other constructive—and both of these agencies were, equally, part of the original order of things. Each of these three philosophical positions had an ancestry going back to before 1700. Each of them handicapped the scientists concerned as well as helping them in their researches. In this respect, too, the evolutionary theory of Charles Darwin was to break the older intellectual moulds, and create a novel pattern of thought.

FURTHER READING AND REFERENCES

The collection, *Forerunners of Darwin, 1745–1859* (ed: Glass, etc.), contains admirable essays on Linnaeus and Maupertuis by Bentley Glass, on Buffon and the situation between 1830 and 1858 by A. O. Lovejoy, on Lamarck by C. C. Gillispie. For general background material on palaeontology, see

John C. Greene: *The Death of Adam*
Loren Eiseley: *Darwin's Century*
Francis C. Haber: *The Age of the World*
C. C. Gillispie: *The Edge of Objectivity*
R. A. Stirton: *Time, Life and Man*

The essays on 'Buffon, Lamarck and Darwin' by J. S. Wilkie and 'Palaeontology and Evolution' in *Darwin's Biological Work* (ed. P. R. Bell) may be strongly recommended. For the development of taxonomy and its relations to palaeontology, the standard works are the two volumes by H. Daudin (in French): see also

E. Guyenot: *Les Sciences de la Vie aux XVIIe & XVIIIe Siècles*
C. E. Raven: *John Ray, Naturalist*

There are good new biographies of Agassiz and Cuvier

Edward Lurie: *Louis Agassiz, a Life in Science*
William Coleman: *Georges Cuvier, Zoologist*

The recent biography, *Erasmus Darwin* by D. King-Hele, claims a more important role for Charles Darwin's grandfather than we are prepared to allow.

9

Life Acquires a Genealogy

DEADLOCKS in science are commonly circumvented, rather than directly broken. While scientists of an older generation grapple brow-to-brow in a head-on confrontation, their juniors can sometimes come to the same problems from a new direction, and so succeed in averting a head-on collision. Yet this redirection of thought is not easily accomplished. It calls for a special kind of originality—and for a man who, having detached himself from the problems and positions of his elders, is free to concentrate on his own novel ideas unhampered by their quandaries.

To such a man, a certain lack of formal qualifications may itself be something of an advantage, saving him from the preconceptions of earlier scientists. So at any rate it seems in the case of Charles Darwin. At a time when scientific research was becoming established as a profession with its own techniques and skills (the English word 'scientist' dates from the year 1840), Darwin began and remained, by both experience and temperament, a gentleman-naturalist. The accepted channels of entry into scientific work were two-fold: mathematics, which included the physical sciences, and medicine, which embraced also physiology. Darwin mastered neither. The love of the countryside which he inherited from his mother and developed in his youth never left him; the standard classical education bored him; the dissecting-theatres at Edinburgh turned his stomach. Putting aside all idea of following his father and grandfather into medicine, he inclined towards Holy Orders, and spent some years ostensibly training for the ministry. Yet, once he had got to Cambridge, his conversations with local naturalists and scientists distracted him from theology as effectively as they had done from medicine earlier, and much of the time that remained he passed on horseback.

The contrast between Darwin's obvious abilities and his failure to apply them tried his father's patience sorely: what was Charles going to make of himself? He never formally renounced his intention to enter the Church. It merely lapsed with the passage of time. After six years of abortive study for two different professions, neither of which had proved to his taste, it finally became clear that Darwin's interests were those of a naturalist, and nothing else: the problem was to find some occupation in which these interests could be usefully exercised. By a happy chance his teacher, Professor Henslow, had the opportunity to recommend him for an appointment as chief naturalist with the naval survey ship *Beagle*, which was due to sail for South America. At first, Darwin's father opposed the plan, but he finally consented; at any rate, the young man would be out of mischief for several years. Until Darwin met Henslow, he had found geology as boring as classics, but Henslow awoke his interest in the subject by showing him its intimate connections with the natural history of the present epoch. In parting, he advised Darwin to take with him on the voyage a copy of Lyell's newly-published *Principles*—'but on no account [as Darwin reported] to accept the views therein advocated'. H.M.S. *Beagle* finally sailed on the 27th December 1831 with Charles Darwin on board. He was then twenty-three years old.

Where Darwin's father saw aimlessness and lack of concentration we can recognize an original mind searching for its proper object. When Darwin left England on the *Beagle*, he had still not found his problem, but in a roundabout way he had prepared himself to meet it. Having been grounded in the classics rather than in mathematics, his grasp of the physical sciences was never very firm, and his distaste for medicine left him without the knowledge of physiology and microscopy which he would certainly have acquired in France or Germany. As a result, though he was much of an age with Claude Bernard, Schwann and Virchow, he was never drawn to the laboratory or the microscope: his grasp of the theoretical sciences was always secondary, and subservient to his obsession with natural history. Later in life, he read what he needed about embryology and the cell theory, but his general command of physiology was only that of a layman. In any event, he read foreign languages—especially German—with difficulty, and he borrowed most of his references to continental scientists from friends or from translations. It was indeed fortunate for Darwin that he met Henslow when he did, for a knowledge of geology was what he needed to give his interests some weight and direction. This apart, his scientific training had been entirely informal: its foundations were laid, not in the lecture-theatre or the laboratory, but among the woods, farms and fields of the English countryside, chatting with stock-breeders, tramping across the fields with his gun, or pausing during a hunt to notice the uncountable seeds of thistledown being carried down-wind. Thanks to the *Beagle*, the training begun in England was to be completed in South America and the Galapagos Islands; and, once there, Darwin could face his central problems directly and candidly, undistracted by received ideas about physiology, embryology or genetics. As things were to turn out, he was as lucky in his ignorance as he was in his knowledge.

Darwin Recognizes His Problem

The *Beagle* was gone nearly five years, returning to Falmouth early in October 1836. Darwin had set out with a conventional belief in the fixity of species, first acquired from the common sense of his time, and reinforced by Henslow, who was an orthodox Catastrophist. His reading of Lyell soon turned him into a Uniformitarian. When the ship sailed, he already had the first part of the *Principles*; and the second volume, including the chapters on zoology, reached him by post at Montevideo in November 1832. From that time on, Darwin was to look at living nature with the same eye that Lyell had brought to geology—an eye for continuities and likenesses, both between present species and their fossil predecessors. At last he had found a task worthy of his intellect; to discover an explanation of the 'succession of organic types', both in space and in time, as natural and 'uniformitarian' as that which Lyell had given to geological development. As he was to emphasize again and again, 'geographical distribution and geological relations of extinct terrestrial inhabitants of South America first led me to the subject'. Travelling southwards down the Argentine coastline, he was impressed 'by the manner in which closely allied animals replace one another'—just as though they were not entirely distinct populations, but rather so many varieties derived from a common original stock. As the environment varied, so too (he remarked) the same structure might be put to different uses—the wing used by most birds for flying served the ostrich as a sail, and the penguin as a flipper. As for South American geology: the giant fossil armadillos recovered from the Pampas showed extraordinary anatomical resemblances to present-day species barely one-tenth their size.

So Darwin reached the Galapagos Islands, and there he was faced by the last essential element in his problem. Up to that point, the general pattern had been one of continuity alone, a steady variation in organic forms coinciding with gradual changes in the environment. This had inclined Darwin, like Buffon and Lamarck before him, to look for some mechanism by which the environment itself might transform a single original population into a dozen different species. But now the problem turned out to be one degree more complex. The various islands of the Galapagos archipelago, a score of tiny volcanic slag-heaps grouped together in the ocean six hundred miles west of Ecuador, formed as constant a collection of micro-environments as could reasonably be imagined: all were subject to the same physical influences, all were composed of the same black lava, many were actually within sight of one another. Yet each separate island (it appeared) had its own distinct flora and fauna. The vast and clumsy tortoises varied slightly from island to island, and local inhabitants could tell at sight from which one any specimen came. The plants varied similarly, while, most striking of all, finches of clearly distinct species were collected in different parts of the group—birds with quite different habits, diets and forms, some having sharp and narrow beaks, others curved and stubby ones (Plate 7).

These inter-island variations were so contrary to Darwin's expectations that he was entirely unprepared for them. To begin with, he had collected specimens indiscriminately, without noting their islands of origin—

> I never dreamed that islands, of about fifty or sixty miles apart, and most of them in sight of each other, formed of precisely the same rocks, placed under a quite similar climate, rising to nearly equal height, would have been differently tenanted.

Yet, once his attention was drawn to the variations, he was quick enough to recognize their significance, and from then on he took care to label his specimens accordingly. As he was never to forget, identical physical conditions could provide a setting in which a dozen distinct species might be formed, whose differences could not possibly be explained by climate and geography alone. If he was to substantiate his dawning conviction that existing species represent the present generations of an historical family-tree, he must work out some new theory which would explain, at one and the same time, both how species close in time and space could be so *different*, and how other species widely separated geographically and geologically could be so *similar*.

Having left the Galapagos, *Beagle* completed her circumnavigation, going by way of Tahiti, New Zealand, Australia and Mauritius. En route, Darwin studied the coral formations of the Cocos Islands in the Indian Ocean, where he formed original (and largely sound) ideas about the processes by which coral islands come into existence. These last observations by themselves would have guaranteed him a place in nineteenth-century science, but nothing could drive out of his mind the overwhelming impact of his South American observations. Arriving back in England a year after leaving the Galapagos, he was still haunted by the memory of those finches and turtles, and his mind still buzzed with the questions raised by his experiences. From this time on, he was entirely single-minded.

Darwin was now twenty-seven. He had forty-five years to live, and this time was divided almost exactly in two by the publication of his *Origin of Species* in 1859. Yet the rest of his life-story is quickly told. He lived for a short while in London, where he made the personal acquaintance of Lyell, and created a great impression at the Geological Society by his paper on coral formations. But he found the city 'a vile smoky place'; and, after his marriage, he retired to the village of Downe, near

Sevenoaks in Kent, to settle for good in the house which still exists today as he left it. Installed there, he worked with an absolute concentration unknown in his student days. He had found his mission, and he was determined to fulfil it. He could exorcize his father's exacting requirements, and pursue his enquiries as a naturalist whole-heartedly. His wife protected him from interruptions. He acquired—perhaps even cultivated—a reputation for ill-health which allowed him to evade the tyrannous demands of Victorian family, society and professional life with impunity. (One recalls Florence Nightingale, that supreme nineteenth-century hypochondriac, who after returning from the Crimea continued to rule the Army Medical Services from her sick-bed up to the age of ninety.) The conflict of obligations at the root of Darwin's anxieties was genuine enough. His self-imposed intellectual demands could never be reconciled with the external requirements of his social position. If his anxieties became neurotic, that is understandable: fortunately for us, the conflict—or chronic illness—disabled him only socially, while he found his escape in creative scientific work. From 1836 on, he never stopped following the trains of thought initiated in the voyage of H.M.S. *Beagle* and, almost without exception, his published works relate back to the field-work which he did in the course of that journey round the world.

Within the framework of this life-story, Darwin's ideas gradually elaborated and clarified. Almost at once, he was compelled to acknowledge the *fact* of organic evolution—

> On my return home in the autumn of 1836, I immediately began to prepare my Journal for publication, and then saw how many facts indicated the common descent of species.

For the time being, he was at a loss to understand how this evolution by descent came about. So in July 1837 he started keeping notes on the 'transmutation of species' in which, as he put it later—

> I worked on true Baconian principles, and without any theory collected facts on a wholesale scale . . . facts which bore in any way on the variation of animals and plants under domestication and nature.

For more than a year he worked solidly, bringing together relevant observations both from his reading and from his first-hand experience. De Candolle, a Swiss botanist, had already noted the importance of competition between species, and Lyell had quoted his observations in the 1830s—

> All the plants of a given country are at war with one another. The first which establish themselves by chance in a particular spot tend, by

> the mere occupancy of space, to exclude other species—the greater choke the smaller; the longest livers replace those which last for a shorter period; the more prolific gradually make themselves masters of the ground, which species multiplying more slowly would otherwise fill.

Lyell had generalized this 'war between the plants', arguing for a 'universal struggle for existence', in which 'the right of the strongest eventually prevails; and the strength and durability of the race depends mainly on its prolificness'—

> Every species which has spread itself from a small point over a wide area must . . . have marked its progress by the diminution or the entire extirpation of some other, and must maintain its ground by a successful struggle against the encroachments of other plants and animals.

Yet Lyell had seen this 'extirpation' only as an instrument by which some species were eliminated: the resulting selection had the power to *destroy existing species*. Darwin's genius was to see how that same selection might *create new species*—a process which for Lyell had been entirely mysterious.

As Darwin soon discovered, there were two separate aspects to his fundamental problem. He must explain both how new variant forms of animal and plant appeared in the first place, and also why certain of them were able to establish themselves at the expense of their rivals. Lyell's idea of the 'struggle for existence' threw some light on the second of these questions, and this particular side of Darwin's theory crystallized out as early as October 1838, when he read Malthus' *Essay on the Principle of Population*. Malthus' pioneer excursion into population-theory had originally been published in 1795, as an economic sermon on the vanity of human wishes. He argued that all schemes for perfecting the working of human society were doomed to shipwreck on one inescapable fact—as time goes on, human population continually tends to increase 'geometrically' (like compound interest), whereas the means of life increase at best 'arithmetically' (like simple interest). Whatever human beings do, he concluded, population will always outrun the food-supply. Any measure to increase social welfare will in the long run enhance fertility, and so sharpen the pressure of population on resources: liberal-minded efforts to better society must inevitably bring nearer the misery, starvation or war by which the excess population would eventually be eliminated—and these evils could be avoided (he added, on second thoughts) only by an unlikely degree of 'moral restraint'. Like Vico in his darker moments, Malthus was gloomily convinced that the conditions of social existence had been established by Providence, and that those who attempted to step outside them brought upon themselves a pre-ordained retribution.

Darwin was not concerned with the social message of Malthus' *Essay*. What caught his attention was the precision with which Malthus had described the effects of competition for the means of survival—the same phenomena that de Candolle and Lyell had remarked on among plants and animals. Darwin did not learn anything entirely new from Malthus; but the vigour of Malthus' argument turned the struggle for the means of subsistence into the focus around which Darwin organized his thoughts about the survival of novel forms. As he put it:

> Being well prepared to appreciate the struggle for existence which everywhere goes on from long-continued observation of the habits of animals and plants, it at once struck me [on reading Malthus] that under these circumstances favourable variations would tend to be preserved, and unfavourable ones to be destroyed. The result of this would be the formation of new species. Here then I had at last got a theory by which to work.

This autobiographical report is of course highly condensed. By the end of 1838, Darwin had come to see that, *if* variant forms capable of being inherited by later generations occurred sufficiently often, and *if* such variant individuals lived in an environment where that particular variation enhanced their chances of multiplying at the expense of their fellows (without the effects of the variation being submerged in cross-breeding), *then* a Malthusian theory would explain why the animals actually surviving the struggle for existence would consist predominantly of the 'best-adapted' forms. Furthermore, given *sufficient time*, populations originating from the same parent-groups, but exposed to different physical conditions and competitors, might end with characteristics quite as different from one another as those produced by the deliberate contrivance of animal-breeders and pigeon-fanciers. This was the mechanism Darwin christened 'natural selection'—a name he chose deliberately so as to contrast it with the 'artificial selection' practised by stock-breeders.

The significance of this new idea lay in the double action of the selective process. Any animal or plant population was exposed, on this theory, to two sets of factors acting together: the *physical* environment—climate, soil and so on—and the *biological* environment—food supply, predators and competitors. The joint action of these two complementary environments at last made sense of his South American observations. The smooth gradation of organic forms (fossil and living) on the mainland reflected the coarser selective action of changing climate and habitat over vast periods of time: the more detailed differences between the finches and tortoises of the various Galapagos Islands reflected the more delicate biological balance, which had favoured seed-eating birds on one island, insect-eating birds on another. His confidence bolstered by the discovery of Malthus' argument, Darwin was now ready to develop his theory at full length.

The Creation of the 'Origin'

Darwin spent five years working out the lines of thought which provided the skeleton of his *Origin of Species*. Most of his fundamental insights had already come to him by the end of 1839, but he continued to collect further evidence and illustrations of his central thesis, and the sheer weight of these examples was eventually to give the *Origin* much of its tremendous intellectual force. By mid-1842, he was ready to produce the first brief abstract of his theory, some thirty-five pages long: he kept this for his own private use, and it came to light among his effects only in 1896. At this stage, evidently, the heart of his argument lay in the analogy between selective breeding among domestic animals and the supposed effects of natural competition in the wild.

> I. The Principles of Var.[iation] in domestic organisms.
>
> II. The possible and probable application of these same principles to wild animals and consequently the possible and probable production of wild races, analogous to the domestic ones of plants and animals.
>
> III. The reasons for and against believing that such races have really been produced, forming what are called species.

By 1844, this first summary had grown to some 231 pages, by which time—one might have thought—Darwin should have been well on the way to preparing for publication. Yet it was to be another fifteen years before the *Origin of Species* reached the public. Like its great forerunner, Newton's *Principia*, the *Origin* was long in gestation, and an external stimulus was finally required to bring it to birth. In 1858, Darwin received a letter from one of his regular biological correspondents, Alfred Russell Wallace, which put forward an essentially identical theory to the one on which he had been working for so long. Wallace was a botanist, specializing in the orchids of the Malayan archipelago. There he had encountered the same paradoxical combination of likenesses and differences that had faced Darwin in the Galapagos. Though Wallace knew of Darwin's work on the *Beagle*, and exchanged many letters with him on biological subjects, it is certain that Darwin had never fully opened his mind to him about the mechansims of evolution. But Wallace, too, had read Malthus, and the idea of 'natural selection' now occurred to him independently. So events forced Darwin's hand. By a very just agreement, the two men presented joint preliminary papers to the Linneaan Society, and Darwin immediately set to work completing a fully worked-out exposition of his ideas. The first edition of the *Origin of Species* was ready for the printer in the middle of 1859.

Why had Darwin delayed for so long? Was he unready to publish earlier, or merely unwilling? Were his reasons primarily intellectual ones, or did they spring rather from his retiring temperament? As in the case of Newton's *Principia*, factors of both kinds seem to have played a part. For a start, the year in which he wrote his second, 200-page draft (1844) was also the year in which Chambers' *Vestiges of Creation* appeared, and the outcry which greeted that book convinced Darwin that his fellow-countrymen were not yet prepared to give evolutionary ideas a serious hearing. For the *Vestiges* was far from being negligible. Its unnamed author failed, like Lamarck before him, in his attempt to explain how new species develop out of, and supersede, older ones, but he demonstrated clearly how the discoveries of Smith and Cuvier could be adapted to support Lamarck's speculative theory of organic descent. Indeed, the very strength of Chambers' argument helps to explain the virulence of his critics—including Thomas Henry Huxley, who was later to change his mind and become Darwin's chief public advocate. Even the anonymity of the work helped to propagate its scandalous doctrines more widely: some people whispered that the author was actually Albert, the Prince Consort. Yet maybe the *Vestiges* did something to draw the fire, and so clear an eventual passage for Darwin. During the years from 1844 to 1859, men had a chance to think seriously about the outstanding difficulties involved in the theory of evolution. For the moment, however, Darwin took the uproar to heart. A half-baked account of his theory would be worse than useless: it would be damaging. Before risking publication, he must bring together as closely-reasoned and fully-documented a body of arguments as he could prepare.

For he did still have some real difficulties to face. By 1844 he understood how, through the 'survival of the fittest' in the 'struggle for existence', natural selection could radically alter the numerical balance between competing populations of animals and plants. But there remained other profound problems, some of which he never wholly solved: notably, the actual extent of variability in Nature, and the extent to which new favourable variations appearing in the wild could survive cross-breeding for long enough to establish themselves, and so prove their selective advantages. Everyone had always accepted, as a matter of common observation, the fact that individuals of any given species differed slightly. Whether there was any absolute limit to these variations in Nature was harder to decide. Lyell had conceded that no *a priori* limit could be placed on variability, yet even he was inclined to believe that it was in practice limited. Darwin, too, though struck by the enormous changes that animal-breeders had produced, could never wholly convince himself that the rate of variation in the wild approached the rate to be observed in the 'unnatural' conditions of domestication. (Wallace had no such hesitations.) Again, it took Darwin some time to appreciate fully how different physical and biological environments could create positive opportunities—what we now call 'ecological niches'—for new variant forms to exploit, and so could encourage divergence and adaptive radiation.

Lacking effective theories of inheritance and variation, which were still in the future, Darwin could do only one thing—fall back on the unquestionable facts of variation in domesticated animals and plants, and the power of human beings to alter the forms of living species without diminishing their fertility. (The idea that all hybrids and other artificial forms were infertile died hard.) While keeping his notes for a later book on genetics on one side, Darwin confined himself for the moment to producing an argument based on certain provisional assumptions about heredity and variation. And, when Wallace's letter reached him in 1858, he at last faced the necessity of publication, and launched the *Origin* on to the world.

The Structure of the 'Origin'

The book was worth waiting for. Twenty years of preparation had been well spent, and the end-product was a classic of both English style and intellectual architecture. Some far-reaching scientific ideas have crept into the world half formed, or disguised by the ponderous language of learned journals. Darwin's theory, that natural selection resulted in 'the preservation of favoured races in the struggle for life', started with no such disadvantage. By 1858 he had worked out, as fully as anyone could then have done, the interlocking trains of thought which meshed together to provide the basic framework of his theory. He had also had time to polish the roughnesses away from his expression: not that he aimed at literary elegance for its own sake, but simply at that unostentatious clarity through whose sheer efficiency style and thought are exactly matched. A treatise of such scale and novelty might easily have become turgid or pretentious; yet Darwin's exposition remained lucid and modest throughout, and in this way its intellectual force had the fullest effect. Here was something which could not, like *The Vestiges of Creation*, be swept aside as an amateur effort. Taken individually, Darwin's arguments were so cogent and simple that they cried out for acceptance: taken together, they enforced conclusions which had hitherto seemed outrageous. The issue of evolution could not longer be burked.

The *Origin of Species* was, accordingly, a decisive document in the development of scientific thought, and however much Darwin's absolute originality is called in question it will always keep this historical significance. Once again, the resemblance to Newton's *Principia* is striking. In each case, one can unravel the complete argument into a dozen strands, all of which had to some extent been anticipated by earlier writers, but this fact does not diminish the magisterial character of the final work. For the idea that great science is a kind of intellectual exploration, whose merit comes simply from planting flags in new territories, is largely a fiction. Originality is not enough. Irreversible changes in scientific thought come about, rather, when originality is allied with great breadth and force of mind, so that the full implications of some novel idea are explored, with all the power of careful reasoning, through the widest possible area of subject-matter.

The breadth and cogency of its argument makes Darwin's *Origin of Species* a pleasure and fascination to read, more than a century after its composition. Yet this also makes it almost impossible to summarize: all we can usefully do here is sample the abstracts at the end of each of Darwin's chapters, so as to indicate some of the strands in his argument, and the candour with which he admitted the difficulties still unsolved. He begins with three chapters on the evidence about variation, both under domestication and in Nature, and on the 'struggle for existence' already hinted at by Lyell and Malthus, and then uses these to explain his central concept of 'natural selection'. The action of natural selection (he concedes) is an hypothesis, whose plausibility depends on certain unproved assumptions. All he can do is state these assumptions clearly, and present such evidence as he can quote to make them reasonable.

> *If* under changing conditions of life organic beings present individual differences in almost every part of their structure, and this cannot be disputed; *if* there be, owing to their geometrical rate of increase, a severe struggle for life at some age, season, or year, and this certainly cannot be disputed; then, considering the infinite complexity of the relations of all organic beings to each other and to their conditions of life, causing an infinite diversity in structure, constitution and habits, to be advantageous to them, it would be a most extraordinary fact if no variations had ever occurred useful to each being's own welfare, in the same manner as so many variations have occurred useful to man. But *if* variations useful to any organic being *ever do* occur, assuredly individuals thus characterized will have the best

> chance of being preserved for the struggle for life; and from the strong principle of inheritance, these will tend to produce offspring similarly characterized. This principle of preservation, or the survival of the fittest, I have called Natural Selection. It leads to the improvement of each creature in relation to its organic and inorganic conditions of life; and consequently, in most cases, to what must be regarded as an advance in organization.

As with any hypothesis, of course, this novel concept must stand or fall by its power to explain the actual facts of Nature:

> Whether natural selection has really thus acted in adapting the various forms of life to their several conditions and stations, must be judged by the general tenor and balance of evidence given in the following chapters. But we have already seen how it entails extinction; and how largely extinction has acted in the world's history, geology plainly declares. Natural selection also leads to divergence of character; for the more organic beings diverge in structure, habits, and constitution, by so much the more can a large number be supported on the area,—of which we see proof by looking to the inhabitants of any small spot, and to the productions naturalized in foreign lands. . . .
>
> On these principles, the nature of the affinities, and the generally well-defined distinctions between the innumerable organic beings in each class throughout the world, may be explained. It is a truly wonderful fact—the wonder of which we are apt to overlook from familiarity—that all animals and all plants throughout all time and space should be related to each other in groups, subordinate to groups, in the manner which we everywhere behold—namely, varieties of the same species most closely related, species of the same genus less closely and unequally related, forming sections and sub-genera, species of distinct genera much less closely related, and genera related in different degrees, forming sub-families, families, orders, sub-classes and classes. The several subordinate groups in any class cannot be ranked in a single file, but seem clustered round points, and these round other points, and so on in almost endless cycles. If species had been independently created, no explanation would have been possible of this kind of classification; but it is explained through inheritance and the complex action of natural selection, entailing extinction and divergence of character.

To close the chapter, Darwin adds one long last paragraph in which he employs a vivid and striking metaphor to re-emphasize his central thesis, and this image grips the mind long after the details of the book have been forgotten:

> The affinities of all the beings of the same class have sometimes been represented by a great tree. I believe this simile largely speaks the truth. The green and budding twigs may represent existing species; and those produced during former years may represent the long succession of extinct species. At each period of growth all the growing twigs have tried to branch out on all sides, and to overtop and kill the surrounding twigs and branches, in the same manner as species and groups of species have at all times overmastered other species in the great battle for life. The limbs divided into great branches, and these into lesser and lesser branches, were themselves once, when the tree was young, budding twigs; and this connection of the former and present buds by ramifying branches may well represent the classification of all extinct and living species in groups subordinate to groups. Of the many twigs which flourished when the tree was a mere bush only two or three, now grown into great branches, yet survive and bear the other branches; so with the species which lived during long-past geological periods, very few have left living and modified descendents. . . . As buds give rise by growth to fresh buds, and these, if vigorous, branch out and overtop on all sides many a feebler branch, so by generation I believe it has been with the great Tree of Life, which fills with its dead and broken branches the crust of the earth, and covers the surface with its ever-branching and beautiful ramifications.

Yet before Darwin tackled his main task, of applying the idea of natural selection to explain the facts of geographical distribution, geological succession, anatomical similarity, animal instinct and embryology, certain difficulties had to be acknowledged. The theory of heredity remained in an unsatisfactory state:

> Our ignorance of the laws of variation is profound. Not in one case out of a hundred can we pretend to assign any reason why this or that part has varied. But whenever we have the means of instituting a comparison, the same laws appear to have acted in producing the lesser differences between varieties of the same species, and the greater differences between species of the same genus. Changed conditions generally induce mere fluctuating variability, but sometimes they cause direct and definite effects; and these may become strongly marked in the course of time, though we have not sufficient evidence on this head. Habit in producing constitutional peculiarities and use

> in strengthening and disuse in weakening and diminishing organs, appear in many cases to have been potent in their effects. . . .
>
> Whatever the cause may be of each slight difference between the offspring and their parents—and a cause for each must exist—we have reason to believe that it is the steady accumulation of beneficial differences which has given rise to all the more important modifications of structure in relation to the habits of each species.

As with 'the cause of gravitation' in Newton's theory, the cause of variation was something which Darwin was compelled to put aside for later study. Like the existence of gravitation, the occurrence of inheritable variations must be accepted for the moment as a simple hypothesis, whose consequences and implications were to be explored.

In the second half of the *Origin*, Darwin examined these implications. Now at last he could show how the idea of natural selection would make new, and very good sense of the geological and geographical facts. There were residual difficulties even in this area, such as the rarity of fossil remains representing transitional forms between species of the better-known epochs, and these compelled him to make yet one more assumption—that the surviving rocks of the Earth's crust were an even more fragmentary set of samples of the actual geological past than Lyell had implied. But once again the assumption might be justified by its consequences:

> I have attempted to show that the geological record is extremely imperfect; that only a small portion of the globe has been geologically explored with care; that only certain classes of organic beings have been largely preserved in a fossil state; that the number both of specimens and of species, preserved in our museums, is absolutely as nothing compared with the number of generations which must have passed away even during a single formation; that, owing to subsidence being almost necessary for the accumulation of deposits rich in fossil species of many kinds, and thick enough to outlast future degradation, great intervals of time must have elapsed between most of our successive formations [etc., etc.] These causes, taken conjointly, will to a large extent explain why—though we do find many links—we do not find interminable varieties, connecting together all extinct and existing forms by the finest graduated steps. It should also be constantly borne in mind that any linking variety between two forms, which might be found, would be ranked, unless the whole chain be perfectly restored, as a new and distinct species; for it is not pretended that we have any sure criterion by which species and varieties can be discriminated.

> He who rejects this view of the imperfection of the geological record, will rightly reject the whole theory. For he may ask in vain where are the numberless transitional links which must formerly have connected the closely allied or representative species, found in the successive stages of the great formation?

These difficulties apart, 'the theory of descent with modification through variation and natural selection' led to a consistent and convincing interpretation of the facts of geographical distribution and palaeontology. To cite only one example:

> On these same principles we can understand, as I have endeavoured to show, why oceanic islands should have few inhabitants, but that of these, a large proportion should be endemic or peculiar; and why, in relation to the means of migration, one group of beings should have all its species peculiar, and another group, even within the same class, should have all its species the same with those in an adjoining quarter of the world.
>
> We can see why whole groups of organisms, as batrachians and terrestrial mammals, should be absent from oceanic islands, whilst the most isolated islands should possess their own peculiar species of aerial mammals or bats. We can see why, in islands, there should be some relation between the presence of mammals, in a more or less modified condition, and the depth of the sea between such islands and the mainland.

So the juggernaut of Darwin's argument rolled on. Nowhere did he pretend to prove things that he could not prove: he was content to present the plausible assumptions on which his conclusions were based for what they were. By the end of the *Origin* he had built up neither a collection of novel discoveries nor an Euclidean system of rigorous demonstrations, but rather a network of *intelligible relationships*. From now on, men could interpret the worlds of geology, palaeontology and zoology in a new, *historical* way. Many things which had hitherto seemed miraculous proved to be 'only natural': the organic world, too, could be brought within the system of uniform forces and causes which Lyell had applied in geology.

In one important respect, indeed, Darwin's theory was more genuinely historical even than Lyell's. Uniformitarian geology had assumed an unending balance between the constructive and destructive agencies shaping the Earth's crust: on to this balance, Darwin superimposed a new, progressive element, associated with the gradual development of organic species. If his assumptions proved well founded, the deadlock between Uniformitarians and the Catastrophists could thus be broken, and Hutton's eternal sequence of geological cycles—lacking dicoverable beginning or end—could be reinterpreted as a continuing process of sequential development. The qualification remained—*if* his assumptions proved well founded—and whether they would do was not at once obvious. For the moment, Darwin was like a cryptographer who claims to have broken a hitherto-undecipherable code. The principles of his decipherment were so many hostages to fortune, and in 1860 their ultimate fate was still uncertain. But this much was certain: whatever in Darwin's theory had to be abandoned, supplemented or modified, this was the first thorough-going account of the evolution of species by 'descent with modification'. If Darwin's theory made good its claims, the temporal scope of natural science would be greatly extended, and the history of Nature would become authentically developmental. If it failed, then at any rate no alternative decipherment was yet in view.

The Scientific Objections

As Darwin foresaw, the *Origin of Species* aroused bitter feelings and triggered off a new outburst of controversy. In the resulting atmosphere of indignation and recrimination, it was not easy to keep the scientific issues in proportion, or to discuss the theory on its zoological merits, apart from questions about its wider repercussions. In consequence, those who had external reasons for wishing to rebut the new doctrine tended to exaggerate the technical objections, flogging the genuine difficulties still facing the theory of evolution for more than they were worth: meanwhile, Darwin's henchmen were tempted in return to play down these difficulties in public, for fear of giving prejudiced objectors too much of a foothold.

From a hundred years' distance, however, it is at last possible to separate the scientific objections to Darwin's theory from the prejudices which it had to overcome. This is worth doing, both to clarify the actual content of Darwin's ideas, and also to do justice to those of his contemporaries —notably Lyell—who were not immediately convinced by his argument of 1859. For there where at least nine distinct lines of counter-argument, some based on logical considerations, others on points of natural history, physiology, genetics, geology and even thermodynamics. None of them was wholly without basis: a few were extremely powerful, and caused serious anxiety to Darwin himself.

Logical Objections

We may start with two criticisms which rested on points of logic. Opponents of evolution have at all times questioned whether the arguments advanced in support of the Darwinian theory amount, in any sense, to 'proofs'. Even with the additions that twentieth-century science can make, the evidence on which the theory of evolution by natural selection relies is fragmentary, needs to be interpreted with the help of some further assumptions, and even then (the critics say) only hangs together more or less consistently: by geometrical standards, at any rate, it yields no conclusive demonstration of Darwin's principles. Now one must admit—as Charles Darwin himself did—that there is a good deal of weight behind this objection. Certainly his speculative imagination reached a good way beyond the 'hard data' available in the 1850s. He was indeed *reinterpreting* the established facts to form a new picture of Nature, and at first this existed only in his own head. He had, of course, to satisfy himself that the data actually to hand could be arranged satisfactorily within the boundaries of the new picture, but to begin with many of these facts remained isolated from one another, like a handful of pieces from a much larger jigsaw puzzle. So it is equally unfair to criticize Darwin's opponents for not being immediately convinced, or to condemn Darwin himself for not producing at a single stroke a picture which was complete and verifiable at every point. For the very fact that Darwin's theory *did* hang together was by itself quite an achievement. Here as elsewhere, this coherence was a good sign that the decipherment was generally sound; and Darwin could justifiably point out how many facts that had hitherto been mysterious found an intelligible place in his new picture.

Granted, 'natural selection' was at the outset an hypothesis—and, to that extent, unproved. But this was not the point:

> It can hardly be supposed that a false theory would explain, in so satisfactory a manner as does the theory of natural selection, the several large classes of facts above specified. [This] is a method [of arguing] used in judging of the common events of life, and has often been used by the greatest natural philosophers. The undulatory theory of light has thus been arrived at; and the belief in the revolution of the earth on its own axis was until lately supported by hardly any direct evidence. It is no valid objection that science as yet throws no light on the far higher problem of the essence or origin of life. Who can explain what is the essence of the attraction of gravity? No one now objects to following out the results consequent on this unknown element of

> attraction; notwithstanding that Leibniz formerly accused Newton of introducing 'occult qualities and miracles into philosophy'.

As in all 'code-breaking', once Darwin's interpretations extended their range far enough, their very coherence began to amount to 'proof'. It was no good demanding a completely verified reconstruction of the History of Nature, for this is something which nobody will ever be in a position to give.

A second logical objection was also raised against the idea of evolution, particularly in France, where Cuvier's ideas still remained of great importance. This argument sometimes became caught up in a terminological tangle: the French word *évolution* referred primarily to 'individual development' (what biologists nowadays call 'ontogeny'), while 'evolution' in Darwin's sense was referred to as *transformation* or *transformisme*. This second objection can be found, for instance, in the *Ontologie Universelle* by Cuvier's pupil, Flourens, as part of a polemic against Lamarck and evolution generally. Ever since Ray, Flourens argued, the definition of the term 'species' had entailed that two different species must be *genealogically* distinct: this being so, the theory of *transformisme* could not be stated as a doctrine about 'species' at all—let alone throw light on the origin of species. For the question, 'Do the cat-species and the lion-species share common remote ancestors?' was, strictly speaking, not so much a daring speculation as a *pseudo-question*—a logically inadmissible form of words. ('These so-called common ancestors, to what species did they belong—cats or lions? If they were cats, they cannot have been the ancestors of lions; if they were lions, they cannot have been the ancestors of cats. Either way, the suggestion is contradictory!') Orthodox French zoologists conducted prolonged rearguard action against Darwinism with the help of such semantic arguments, as much as of empirical evidence.

Behind this somewhat perverse dialectic lay a more substantial difficulty. Ray's conception of a 'species' presented a theoretical ideal, rather than a fact of nature, and organic species in this absolute sense could 'have a real existence in nature' only if the theory of descent were rejected. With that English pragmatism which has so often infuriated the more Cartesian French mind, Darwin continued to write about the transformation of species, even in the process of destroying the criterion by which the distinctness of species was guaranteed. To carry apparent inconsistency further: he went on to claim both that all taxonomic classifications rested on arbitrary human decisions and that a classification based on evolution revealed for the first time the true relationships between taxonomic groups. Once again this objection, though not fatal, reminds us just how radical Darwin's theory was, and how far-reaching its intellectual consequences were eventually to be. For Flourens was right in this:

if evolution were accepted as a fact, the theoretical foundations for the notion of a species were undercut. Darwin might go on using the term as an approximation, but he could do so only roughly and intuitively, pending a redefinition of the notion in the light of an evolutionary critique.

In the event, this redefinition has waited until the last few years, forming the subject-matter of the 'new taxonomy' of Mayr, Dobzhansky and Simpson. Their work makes it clear that the older idea of a 'species' can be carried over into an evolutionary zoology only with certain definite restrictions. Suppose, for example, a palaeontologist traces a sequence of fossils through the superimposed layers of a geological formation (e.g. those of the extinct sea-urchin *Micraster*): he will find a mixture of fossil forms at every level, and the proportions of these different forms in the mixture will change gradually as he goes through the sequence of layers. As a result, any decision to say that, between two given adjacent levels, one 'species' ends and another 'species' begins will be completely arbitrary —even though specimens from levels widely separated in time are clearly distinct, and anyone would allocate them to different species. Similar difficulties arise over the classification of organisms which reproduce asexually, e.g. certain kinds of dandelion and earthworm. Only for populations of creatures which reproduce sexually, and inter-breed freely over a continuous area, can one still provide a criterion for distinguishing species with the clarity biologists demanded in the days before Darwin; and, even then, this criterion applies strictly only for one given phase of history.

The Problem of Variation

These taxonomic difficulties were bound up with another problem, which involved the greatest single gap in Darwin's theory. This was his silence about the mechanism of variation and inheritance. Heredity and mutation form the central topics of modern genetics, and in recent years nothing has done more to strengthen the Darwinian theory than the support of genetical discoveries. But this happy alliance has been a twentieth-century development: for the first forty years, the theory of evolution had to make its way independently of the science of inheritance, and in some respects the two sciences were actually at odds.

Darwin recognized all along that, in order to complete the theory of evolution by variation and natural selection, a satisfactory account must be given of the mechanism of reproduction and inheritance. But since he was never much of a physiologist, his ideas about the subject—published in his book, *The Variation of Animals and Plants under Domestication* (1868) —had a distinctly amateur tone, as compared with those of a professional like August Weismann. In the *Origin*, Darwin had carefully sidestepped questions about inheritance, since these would only have added further confusion to the complex zoological questions with which he was directly concerned. To account for the persistence of certain novel variations, he merely invoked 'the strong principle of inheritance'; but he did nothing to indicate why some variations were inherited, while others were transitory, or how the characters of parents determined those of their offspring. In his book on *Variation*, he at last set out to explain these things, but the resulting theory had only a very short life: its chief value was that it provoked the critical re-examination of the subject by which Weismann started genetical thought developing along its modern track.

The fundamental units in Darwin's theory were minute particles called 'gemmules'. These were continually produced in every part of the adult body and passed to the sex organs, to be incorporated into the spermatozoa and ova. The original cells from which the embryo developed accordingly reflected the condition of the parents at the time of conception: as a result, characteristics acquired during the parents' lifetime could be transmitted to their children. This meant that the environment directly influenced the character of offspring, and so the course of evolution: a temporary pigmentation could this way become a permanent physiological feature of a species. In the twentieth century, this form of theory is usually associated with Lamarck's name rather than Darwin's, and the idea of evolution by natural selection and Mendelian inheritance ('neo-Darwinism') has been sharply opposed to any 'Lamarckist' theory of evolution involving the inheritance of acquired characteristics. Yet this opposition was created only by Weismann's critique and the work it stimulated. In Darwin's own mind, there was no such rivalry. As time went on, indeed, he was inclined to yield more and more ground to Lamarck and the inheritance of acquired characteristics. In the definitive edition of the *Origin* (1872), he expressly emphasized the 'non-Darwinian' elements in his theory:

> I have now recapitulated the facts and considerations which have thoroughly convinced me that species have been modified during a long course of descent. This has been effected chiefly through the natural selection of numerous successive, slight, favourable variations; aided in an important manner by the inherited effects of the use and disuse of parts; and in an unimportant manner, that is in relation to adaptive structures, whether past or present, by the direct action of external conditions, and by variations which seem to us in our ignorance to arise spontaneously. It appears that I formerly underrated the frequency and value of these latter forms of variation, as leading to permanent modifications of structure independently of natural selection.

By now, he was almost apologetic about his great discovery:

> But as my conclusions have lately been much misrepresented, and it has been stated that I attribute the modification of species exclusively to natural selection, I may be permitted to remark that in the first edition of this work, and subsequently, I placed in a most conspicuous position—namely, at the close of the Introduction—the following words: 'I am convinced that natural selection has been the main but not the exclusive means of modification.' This has been of no avail. Great is the power of steady misrepresentation; but the history of science shows that fortunately this power does not long endure.

In the light of Darwin's theory of 'gemmules' or 'pangenes', one can see why he was so open to Lamarckian ideas, for not only was the inheritance of acquired characteristics something entirely possible, it was something to be expected. Pure Darwinism, as men were later to think of it, comprises the ideas of the *Origin* divorced from those of the book on *Variation*, with all references to acquired characteristics eliminated. As we saw in *The Architecture of Matter*, Weismann and his successors demonstrated that the germ-cells from which the embryo grows reflect, not the adult condition of the parents at the time of conception, but rather their own inherited genetic constitution. In this 'continuity of the germ-plasm', Weismann saw the key to the extraordinary continuity of form displayed by organic species. So the influence of the environment on genetic inheritance was reduced to a minimum, and it became necessary to look elsewhere for the ultimate source of inheritable variations, or 'mutations'.

Modern genetics has also resolved two further difficulties in Darwin's original theory. He had assumed that 'numerous successive, slight, favourable variations' were by themselves enough to produce an unlimited transformation of organic forms by the action of natural selection. The subsequent debate made this assumption appear extremely implausible. If individual mutations were so slight their selective advantages would be correspondingly small, and in all probability their effects would be swamped in subsequent cross-breeding. Gradually, evolutionists became convinced that variation must take place in a 'saltatory' manner, by jumps, some of the resulting modifications having a significant selective effect. This belief has been confirmed since Darwin's death by genetical studies of mutation.

A related difficulty has to do with the *direction* of the variations assumed on Darwin's theory. Though he spoke of natural selection as acting on 'numerous successive, slight, *favourable* variations', it was the essence of his doctrine that variations were, in themselves, neither 'favourable' nor 'unfavourable': they merely *happened*. Their favourable or unfavourable character showed up only subsequently, in the competitive 'struggle for life'. On this point, also, Darwin weakened under attack, giving ground in a Lamarckian direction to an extent which subsequent discoveries have shown to be unnecessary. Here, his colleague Wallace saw more clearly the shape of ideas to come. If variations in the wild were rare and surprising, then—in order to make evolution possible at all—they must occur not randomly but in direct response to the demands of the environment. The opposite was in fact the case. There was no shortage of variant forms in Nature: rather, as Wallace argued in a letter to Darwin (1866)—

> It would be better to do away with all such qualifying expressions and constantly maintain (what I certainly believe to be the fact) that, *variations of every kind are always occurring in every part of every species*, and therefore the favourable variations are *always ready* when wanted. You have, I am sure, abundant materials to prove this, and it is, I believe, the grand fact that renders modification and adaptation to conditions almost always possible.

Yet in this respect, too, Darwin never became a 'pure Darwinian'. Throughout his later years he kept harking back to the thought that variation was somehow unnatural, and occurred more frequently under domestication than in the wild; and that, in order to yield favourable results, the direction of variation must be substantially influenced by the species' current conditions of life. Like so many great scientific innovators, Darwin remained entrammelled in ideas which were later to be entirely superseded by his own novel suggestions.

Embryological Difficulties

Among the most challenging problems, however, were those presented by embryology, for there were certain striking and puzzling facts which Darwin's opponents regarded as inexplicable on any evolutionary view. The whole subject is fascinating and complex, and one cannot do it justice in a brief summary. To put the chief points succinctly, however, one can say this: firstly, the early embryonic forms of different animals from the same general group (e.g. the vertebrates) resemble one another very much more closely than the adult forms of the same species. Secondly these embryonic forms resembled in some respects also the forms of *extinct* species (from which, on Darwin's view, they were descended)—even to the point of developing rudimentary organs, such as gill-slits, which may have no possible function either in the embryo or in the adult life of the species. The parallels between present-day embryos and their supposed predecessors made so deep an impression in the nineteenth century that some enthusiastic evolutionists (such as the German naturalist, Haeckel) elevated it into a general law. 'Ontogeny recapitulates phylogeny'—during development, any embryo passes through a succession of forms which represent its adult ancestors in their evolutionary order.

So stated, the doctrine of recapitulation does not stand up to close scrutiny, but the findings of embryology posed fascinating riddles to zoologists. For Louis Agassiz, who was by now as supreme a zoological pontiff in America as his master Cuvier had been in France, these embryological riddles were one of the principal weapons in a long rearguard action against Darwinism. As he argued, embryos were shielded almost perfectly from the influence of the environment; so natural selection—as he understood it—could scarcely play any part in determining their forms. The theory of natural selection could therefore throw no light on these embryological riddles. If the embryonic forms of terrestrial mammals displayed transient aquatic features, this could have no significance for the 'survival' of those mammals in the 'struggle for life'. Such embryological resemblances and anomalies must find their explanation elsewhere. They were, presumably, evidence of the Great Design of Creation—of the fundamental ideas or blueprints, according to which Deity fashioned organic species afresh after each geological catastrophe.

Agassiz' argument appealed to those who were anxious to maintain a foothold in the organic world for divine intervention. But the form of his objection showed that he had not appreciated the dynamic and historical character of the Darwinian view. Of course, evolutionists replied, not every character of existing species is a consequence of natural selection acting in the *present* conditions of life. The whole point of the Darwinian theory lay in the historical succession of processes which, step by step and over a long period of time, had brought living creatures to their present forms. Once this was recognized, Agassiz' objection could be turned against him. As he rightly said, the embryos of higher animals are shielded from many of the selective pressures that operate freely during post-natal life. Looked at historically, this means that selective influences can do less to *eliminate* functionless forms during these embryonic stages than during the post-natal existence. Just as the overburden of soil protected fossils from the destructive action of wind and weather for millions of years, so the womb screened ancestral embryonic characteristics against the destructive agencies of natural selection, even after these characteristics had lost all positive function. By arguments such as these, the riddles of embryology became in due course some of the most powerful arguments *for* the theory of evolution.

The Demonstration of Natural Selection

This by no means exhausts the tally of difficulties facing the evolutionists. Though by their theory they could relate together a dozen classes of fact which hitherto had been intellectually unrelated, it was nevertheless very desirable to demonstrate these hypothetical agencies in action at the present day. For there was, as men rightly pointed out, no direct evidence whatsoever of the operation of natural selection. This fact alone gave Lyell a reason for hesitation. In all his geological work, he had been careful to introduce into his explanations only those forces and agencies which one could actually see at work: wind, rain, frost, volcanoes and so on. Yet, in the case of natural selection, Darwin had been unable to produce any convincing demonstration of its continuing action. Though his arguments that it *must* operate were powerful—and, in the long run, convincing even to Lyell—positive proof of its operation was lacking.

As things turned out, this was one of the last deficiencies to be made good. For a really satisfactory illustration of natural selection producing observable effects over periods of time accessible to human record, one must turn to very recent studies of insect populations. Certain species of moths, for instance, have patterns on their wings closely resembling the bark of the trees on which they rest: these patterns serve as camouflage, helping to conceal the moths from the birds which prey on them (Plate 9). The range of wing-patterns and colourings is very variable, and in some cases shows a striking division into those patterns which are predominantly light. and those which are predominantly dark. Recently, naturalists have compared the proportions in which the two kinds of colouration appear in different parts of Britain. In the open countryside of North Wales, for example, moths with very dark wing-patterns are rare; but in the adjoining industrial regions, where the trees are blackened by industrial grime, the dark wing-patterns predominate and it is the lighter forms which are rare. Since this blackening of the trees dates back less than two hundred years, it seems certain that in the eighteenth century the moth population over the whole area would have shown a uniform distribution of wing-patterns.

Naturalists regard this as a perfect illustration of the manner in which natural selection, acting in this case through camouflage, leads to the preferential survival of an organic form which was initially rare, to a point at which it becomes predominant. The illustration is striking, but one comment is in place. The dark-winged forms have certainly profited by the selective advantage they possess in industrial areas, but they have not yet become a distinct species. Though we can now point to 'industrial melanism' as an instance of natural section operating today, we have still not shown natural selection producing two distinct species of descendants from one original population in the wild.

Geological Difficulties

Until 1899, another similar objection was brought against evolution-theory. This pointed to the absence of any direct demonstration that there was continuous development of organic forms through earlier geological epochs. As we saw earlier, the strength of the 'catastrophist' view lay in the fact that the fossil record was highly discontinuous. Though there were resemblances between (say) fossil mammoths and present-day elephants, the gaps in the record were very much larger than could result from any genetic mutation, however startling. Anyone who accepted the theory of descent had to argue that the absence of intermediate forms was evidence only that these particular forms had for some reason been lost. This might have happened for a number of reasons: either because during those phases of evolution no geological formations were being laid down, or alternatively because the evolutionary changes were so rapid that the number of intermediate individuals was comparatively small, and the chance of their surviving as fossils correspondingly reduced. Failing any positive illustration of fossil forms changing from one species to another by a gradual and continuous development, evolutionists were inevitably on the defensive.

This positive demonstration came only when A. W. Rowe published a classic report on the fossil sea-urchin *Micraster*, referred to earlier, which is preserved in the chalk-layers of north-west Europe. Rowe worked over the chalk-face of Beachy Head inch by inch, and collected fossil specimens from every layer. Having securely established the temporal succession of some two thousand specimens, he was able to show that here—where the conditions for geological formation had favoured the survival of a representative selection of fossils—no discontinuity was to be found. From the oldest layers to the newest, the forms of the sea-urchins developed smoothly and continuously, just as Darwin's theory required.

Even so, there remained some residual difficulties about the extent of geological time. Darwin had seized on Lyell's analogy, and described the geological record as

> a history of the world imperfectly kept, [of which] we possess the last volume alone . . . [and even of this] only here and there a short chapter has been preserved; and of each page, only here and there a few lines.

But, even among scientists who accepted Lyell's 'geological uniformitarianism', there were many who found the chronological demands of Darwin's 'biological uniformitarianism' excessive. As Darwin was eventually forced to admit:

> That the geological record is imperfect all will admit; but that it is imperfect to the degree required by our theory, few will be inclined to admit . . .
>
> I have felt these difficulties far too heavily during many years to doubt their weight. But it deserves especial notice that the more important objections relate to questions on which we are confessedly ignorant; nor do we know how ignorant we are. We do not know all the possible transitional gradations between the simplest and the most perfect organs; it cannot be pretended that we know all the varied means of Distribution during the long lapse of years, or that we know how imperfect is the Geological Record. Serious as these several objections are, in my judgement they are by no means sufficient to overthrow the theory of descent with subsequent modifications.

Lyell had spoken of the Earth's past only as 'indefinite' in extent. Darwin's demands were rather more definite: they implied a sequence of phases lasting some ten million years apiece. Darwin's conclusion that the surviving geological formations represented in fact only a minute fraction of the Earth's history was the hardest of all for geologists to swallow: in many cases, he declared, the fossil record showed that the time difference between the periods of which adjacent strata were formed was much greater than the lengths of time it had taken to form each of the two strata. Even professional geologists took a long time to accept this doctrine.

The Thermodynamic Time-Scale

The reasons for their hesitation were not solely derived from geology. The last and, as Darwin confessed, 'probably one of the gravest objections as yet advanced' to his theory was Lord Kelvin's thermodynamic argument. Kelvin (Sir William Thomson, as he then was) certainly found the wider implications of Darwin's theory disturbing, and pressed his attack on the theory with some warmth of feeling. Yet, even when one sets aside possible ulterior motives, the difficulties he raised were extremely obstinate. In a series of papers from 1862 on, he revived and brought up-to-date the theory of refrigeration earlier advanced by Buffon and Fourier. Given the classical nineteenth-century laws of heat-production and radiation, one could calculate the rate at which the Earth and the Sun should be cooling through the loss of their own natural heat, and such a calculation set upper limits to the ages of the Earth and Sun consistent with the principles of classical physics. Relying on this theoretical argument, Kelvin worked out that the Sun could not have existed for more than five hundred million years, and that the Earth's loss of heat made it inconceivable that it had supported life for more than a few million years. Hutton and Lyell had been deceived: though Newton and his successors satisfied themselves that the planetary system was *dynamically* stable, they could not at the same time prove that it was *thermodynamically* changeless. The ambition to explain the structure of the Earth's crust by appealing to a permanent balance between constructive and destructive forces had been misconceived, and geologists could no longer make unlimited demands on antiquity. Thermodynamics placed limits on what was possible, and the History of Nature must adapt itself to these limits.

Fortunately, Darwin and his supporters were not irrevocably committed to any absolute estimates of the time required for the development of species and, for a while, they did their best to operate within the narrow limits which Kelvin would allow them. Still, this physical barrier—which they were unable to breach—did much to distort evolutionary theory. In order to accommodate all geological history within Kelvin's time-scale, for instance, T. H. Huxley was compelled to make needless concessions to the catastrophists, and Darwin himself conceded that organic evolution must have proceeded more rapidly in earlier epochs. For forty years, in fact, Kelvin's limitation on the antiquity of life had the same cramping effect on evolutionary thought that the Biblical time-scale had had on Hooke and Steno's geology.

Darwin never saw his way past this particular obstacle: he could only hope that somehow or other it would eventually remove itself. As he wrote in his final revision of the *Origin*:

> With respect to the lapse of time not having been sufficient since our planet was consolidated for the assumed amount of organic change, and this objection, as urged by Sir William Thomson, is probably one of the gravest as yet advanced, I can only say, firstly, that we do not know at what rate species change as measured by years, and secondly, that many philosophers are not as yet willing to admit that we know enough of the constitution of the universe and of the interior of our globe to speculate with safety on its past duration.

For once in a while, natural history proved in the long run more reliable than physical theory. The discovery of radioactivity, more than a dozen years after Darwin's death, showed that all was not known about the sources of physical energy in the world. Nor were the effects of this discovery limited to the theoretical level. In 1909, Joly demonstrated that heat from the radioactive minerals within the Earth could by itself slow down the planet's rate of cooling to such an extent that its probable life was vastly longer than the twenty-four million years of Kelvin's estimate. Joly went further: this 'radiogenic' heat might create thermal instabilities which expanded and buckled the Earth's crust, and so provide the motive-force for mountain-building. By 1931, Kelvin's estimate of the Earth's age had been multiplied by one hundred, and nowadays the accepted date for the condensation is a good five thousand million years back, while the permanent crust apparently formed around 2800 million years ago. Finally, Kelvin's limits on the age of the Sun have been swept away by our understanding of thermonuclear processes: indeed, the fusion of hydrogen atoms to form helium releases so much energy that the Sun may not be cooling down appreciably at all. Thus the theoretical objections which in Darwin's lifetime placed the gravest obstacles in the way of his theory have, since his death, removed themselves as completely as he hoped.

Darwinism and Natural Theology

To most people, however, the technical arguments about evolution were so much irrelevant skirmishing, and did not touch the points of real sensitivity. For accepting Darwinian theory apparently meant abandoning the last strongholds of natural theology in the History of Nature and, this being so, the violence with which some devout scientists reacted is not surprising.

Darwin's old teacher at Cambridge, Adam Sedgwick, viewed the idea of organic development and descent with the same horror that Richard Kirwan had felt for Hutton's *Theory of the Earth*. As he believed, the harmonious balance between organisms and their environments was maintained by periodic interventions of divine Providence. Chambers' *Vestiges* had dispensed with such interventions, and treated the order of living nature as self-sufficient. 'If current in society,' Sedgwick declared, such beliefs could lead to 'nothing but ruin and confusion. . . . It will undermine the whole moral and social fabric, and inevitably will bring discord and deadly mischief in its train.' To his great sorrow, he found similar defects in Darwin's argument, which also destroyed the 'essential link' between the moral and material worlds:

> You have ignored this link; and, if I do not mistake your meaning, you have done your best in one or two pregnant cases to break it.

> Were it possible (which, thank God, it is not) to break it, humanity, in my mind, would suffer damage that might brutalize it, and sink the human race into a lower grade of degradation than any into which it has fallen since its written records tell us of its history.

Sedgwick's reactions were far from unique. Since geologists had discredited the historical tradition of a Universal Flood, the creation—or periodic re-creation—of species had become almost the last illustration to which natural theology could point of the Divine Power manifestly acting in the world of Nature.

Darwin himself tried hard to take the sting out of the religious objections to his theory. In the very first private abstract, written in 1842, he was already including a draft of his eventual peroration:

> It accords with what we know of the law impressed on matter by the Creator, that the creation and extinction of forms, like the birth and death of individuals should be the effect of secondary means. It is derogatory that the Creator of countless systems of worlds should have created each of the myriads of creeping parasites and worms which have swarmed each day of life on land and water on the globe.

Rather, we should recognize that the rationality of the organic world, like the inorganic, lay in its very self-sufficiency. As he wrote in the *Origin*:

> When I view all beings not as special creations, but as the lineal descendants of some few beings which lived long before the first bed of the Cambrian system was deposited, they seem to me to become ennobled.

The closing words of his argument changed only slightly between the first draft of 1842 and the sixth, definitive edition of the *Origin* thirty years later:

> Thus, from the war of nature, famine and death, the most exalted object which we are capable of conceiving, namely, the production of the higher animals, directly follows. There is grandeur in this view of life, with its several powers, having been originally breathed [by the Creator] into a few forms or into one; and that, whilst this planet has gone cycling on according to the fixed law of gravity, from so simple a beginning endless forms most beautiful and most wonderful have been, and are being, evolved.

(This passage suffered one significant alteration: the words 'by the Creator' were inserted during the six weeks between the appearance of the first and second editions.)

Darwin could not, however, avert controversy so easily. He was imposing on men's ideas a change which was too drastic to take place quietly or insensibly. The most obvious result of his argument was to complete the revolution in natural history inaugurated by Hutton in the 1790s, by which the development of the Earth and its inhabitants—first inorganic, and now organic also—joined the Solar System as evidence that the Laws of Nature acted in a constant and consistent manner. Just as Lyell had expelled the Deluge from the geological history of Europe, so now Darwin threw even graver doubt on 'the veracity of Moses as an historian'. Zoology became as much a field for 'secondary causes' as geology, and the second chapter of *Genesis* followed the first, out of history and into allegory. In any case, as his followers argued, the Wisdom of the Creator was better shown in the continued uniform operation of natural causes than by occasional supernatural interventions—however spectacular or catastrophic.

This was the position which the next generation of religious-minded scientists were to take up. For instance, Darwin's leading American supporter (the botanist, Asa Gray) combined a scientific acceptance of natural selection with a sincere belief in Divine Design, and mischievously congratulated Darwin 'for his striking contributions to teleology . . . knowing well that he rejects the idea of design, while all the while he is bringing out the neatest illustrations of it'. In later editions of the *Origin*, Darwin himself was able to quote 'a celebrated author and divine', who had written to him saying that

> He has gradually learnt to see that it is just as noble a conception of the Deity to believe that He created a few original forms capable of self development into other and needful forms, as to believe that He required a fresh act of creation to supply the void caused by the action of His laws.

But this was something which men could, at best, learn to accept only gradually.

Was the religious battle really necessary? By now, it should be clear that it was entirely unavoidable. Adopting the standpoint of twentieth-century orthodoxy, one can no doubt view it all as a sad misunderstanding, and argue that it need never have occurred, if men had only respected the fundamental difference of aim between Science and Religion—*viz*., that 'Science deals with Things, describing their material properties and interactions; Religion deals with Persons, laying down the rules governing inter-personal relations; so that—Science being descriptive, Religion normative—the apparent conflict between them sprang always from a confusion.' As applied to the mid-nineteenth century, however, such a judgement is both too facile and historically irrelevant. For the 'confusion' in question, of basing the rules of conduct—Sedgwick's 'moral order'—on a particular set of beliefs about the History of Nature and Man's place in in it—the 'natural order'—had up till that time been an almost universal element in all systems of religious belief. (Aristotle, Confucius and Epicurus are the only notable exceptions to this generalization.) The majority of men had always seen Man as occupying a unique and central place in Nature, and found the final justification of their ethical and religious conceptions in a cosmic history embracing both Nature and Man. This pattern at any rate was a common element uniting the Myth of Osiris, aboriginal stories of the dream-time, the Hebrew Covenant and the Christian Gospel.

Sedgwick's fears for society might be exaggerated, but he saw quite correctly that Darwin threatened to destroy this pattern. The new theory called in question both the uniqueness of man, and the traditional view of cosmic history. Being an *historical* theory, Darwin's explanation of the origin of species inevitably challenged the historical claims of all alternative, religious accounts of the creation of Man. Here was an issue which deeply affected all Christians—Catholics and Protestants alike; whereas Catholic theologians had to some extent stood aside from the earlier geological debate, they could no longer remain outside the battle when the status of Man became an open question. Man's relationship to God in History was the central theme of the Christian message, and no theologian could happily see all God's works turned into by-products of the blind action of natural selection.

This discomfort was, in fact, felt more widely. It affected not only Christians, but also advocates of the 'secular religions' which were growing up in nineteenth-century Europe. Quite apart from Darwin's zoology, the ideas of historical development, and of progress were the two most significant features of nineteenth-century thought; and—though superficially 'evolutionary' in their implications—both these ideas lent themselves more readily to a 'providential' interpretation. We saw earlier how Herder treated the appearance of Man on Earth as the final fulfilment of a teleological process going on throughout cosmic history—what he called 'the purposes of Mother Nature'. For Lamarck, too, all manifestations of life represented the victory of a constructive element in Nature over the disruptive agencies of the physical world. The secular religions of the late-nineteenth century, Marxism included, derived their fundamental ideas from this earlier, teleological view of Nature, rather than from Darwin's theory of variation and natural selection. The true goals and purposes of human life, they taught, were apparent from the progressive development of Nature and Humanity. Theologically speaking, these secular religions simply carried one stage further the intellectual changes which had already begun two centuries before. Protestant theologians in the seventeenth century had placed the Book of Nature alongside Holy Scripture, and given them both equal authority as sources of religious truth: now, the secular prophets of the nineteenth century made Nature the sole authority for our moral and spiritual conceptions.

The Darwinian theory called in question all teleological interpretations of the History of Nature—theistic and naturalistic alike. It did not deny that organic structure and animal behaviour were adaptive. But it did deny that these functional aspects of Nature came into existence as the end-results of processes specifically aimed at their production. Rather, they were by-products of a variety of interacting factors: over-population, random variation, and the perpetuation of those variations which were advantageous in a particular environment. A mechanism for the creation of new species whose fundamental 'motive-force' was the pressure of population on the means of subsistence offended against the preconceptions both of Marxists and Christians. As a result, George Bernard Shaw could protest, as sincerely as Bishop Wilberforce, against the 'purposelessness' of natural selection:

> If it could be proved that the whole universe had been produced by such Selection, only fools and rascals could bear to live.

A similar motive lies at the base of the Lysenko dispute in our own time. The distaste of Soviet ideologists for twentieth-century 'neo-Darwinism', in which the idea of natural selection is combined with the genetical mechanisms of Mendel and Morgan, rests on something more than agricultural expediency and political prejudice: as they quite rightly see, this combination re-emphasizes the mechanistic element in the Darwinian theory, while the teleological character of the Marxian dialectic harmonizes more naturally with Lamarck's ideas.

On the whole, Christian theologians have extricated themselves from the difficulties into which Darwin plunged them more easily than their 'naturalistic' colleagues. This reappraisal of Christian doctrine has not been without its agonies but, formally, little enough has been required of theology. Philo and Augustine had taught, long ago, that the truth of the Scriptures was as much symbolic as literal, and it did not apparently cost very much to concede that the creation of a· imals and men described in the second chapter of *Genesis* should be understood rather allegorically than historically. By now, it is generally agreed that the crucial historical claims of Christianity—notably, the Resurrection—arise out of the New

Testament story rather than over anything in the *Book of Genesis*. So, although a few fundamentalist sects still reject natural selection—and even the theory of descent—as blasphemous denials of divine Providence, theologians of most denominations have withdrawn gracefully from the positions most directly threatened.

By Darwin's time, then, any remaining idea that God created the World solely and expressly for Man's convenience, had been thoroughly undermined. The new evolutionary perspective stirred Tennyson's imagination to its depths, even in prospect. Having read Chambers' *Vestiges*, he posed in his *In Memoriam* (1850) questions which were beginning to nag at the minds of all educated men:

> Are God and Nature then at strife,
> That Nature lends such evil dreams?
> So careful of the type she seems,
> So careless of the single life;
>
> That I, considering everywhere
> Her secret meaning in her deeds,
> And finding that of fifty seeds
> She often brings but one to bear,
>
> I falter where I firmly trod,
> And falling with my weight of cares
> Upon the great world's altar-stairs
> That slope thro' darkness up to God,
>
> I stretch lame hands of faith, and grope,
> And gather dust and chaff, and call
> To what I feel is Lord of all,
> And faintly trust the larger hope.

Geologists were returning a dusty answer to these questions. On a geological time-scale, species seemed as short-lived as individuals:

> 'So careful of the type?' but no,
> From scarped cliff and quarried stone
> She cries, 'A thousand types are gone:
> I care for nothing, all shall go.
>
> 'Thou makest thine appeal to me:
> I bring to life, I bring to death:
> The spirit does but mean the breath:
> I know no more.' And he, shall he,

Man, her last work, who seem'd so fair, . . .

Who trusted God was love indeed
And love creation's final law—
Tho' nature, red in tooth and claw
With rapine, shriek'd against his creed—. . .
Be blown about the desert dust,
Or seal'ed within the iron hills?

Natural theology was dying painfully, and the sense of loss was extreme. Instead of confirming God's Wisdom, the natural world had transformed itself into a disgusting caricature. Religious faith was still indispensable, but men could no longer find inspiration by contemplating the marvels of the natural world. If the whole story of Man's Creation lay in the struggle for existence—and it is notable how far Tennyson had already grasped the essentials of natural selection, nine years before the publication of Darwin's *Origin*—then, indeed, 'only fools and rascals could bear to live'.

No more? A monster then, a dream,
A discord. Dragons of the prime,
That tear each other in their slime,
Were mellow music match'd with him.

O life as futile, then, as frail!
O for thy voice to soothe and bless!
What hope of answer, or redress?
Behind the veil, behind the veil.

FURTHER READING AND REFERENCES

The centenary of the *Origin of Species* in 1959 stimulated a flood of books on Charles Darwin and his times. The most penetrating biography of Darwin, however, has appeared even more recently: it is that by Sir Gavin de Beer. For a more popular life of Darwin and Huxley, see

William Irvine: *Apes, Angels and Victorians*

The *Origin* itself is available in paperback: the *Voyage of the Beagle* is also well worth reading. Recent anthologies of Darwin's writings are

The Darwin Reader: ed. M. Bates and P. S. Humphrey
C. Darwin and A. R. Wallace: *Evolution by Natural Selection*
C. Darwin (ed. Loewenberg): *Evolution and Natural Selection*

The development of Darwin's ideas, and the intellectual difficulties facing his theory, are clearly analysed in

Loren Eiseley: *Darwin's Century*

For particular aspects of Darwinism, consult

T. A. Goudge: *The Ascent of Life* (philosophical)

A. J. Cain: *Animal Species and their Evolution* (easy introduction to modern taxonomy)

E. Mayr: *Animal Species and Evolution* (the authoritative work on the new taxonomy)

T. Dobzhansky: *Genetics and the Origin of Species*

R. A. Fisher: *The Genetical Theory of Natural Selection*

Gavin de Beer: *Embryos and Ancestors*

G. G. Simpson: *The Meaning of Evolution* and *Tempo and Mode in Evolution*

David Lack: *Darwin's Finches*

W. H. Dowdeswell: *The Mechanism of Evolution* (a good introduction)

See also A. Hunter Dupree's life of *Asa Gray* and David Lack's book on *Evolutionary Theory and Christian Belief.*

10

History and the Human Sciences

FROM the 1860s on, 'Evolution' became a catchword which dominated human thought far beyond the limits of zoology and gave it a characteristic period flavour, just as the ideas of Providence and Order had done in earlier times. But the credit for this fact can go only partly to Charles Darwin, for the changes which he effected in zoology were one aspect of a larger intellectual transformation, which was reshaping men's attitudes towards cosmology and human history as well. The *Origin of Species* may have been the book which (in R. G. Collingwood's words) 'first informed everybody that the old idea of Nature as a static system had been abandoned', and whose effect was 'vastly to increase the prestige of historical thought'. But it could make this impact only because it broke down the artificial barrier between Science—which had hitherto been concerned with the static Order of Nature—and History, which studied the development of humanity. So the two most powerful intellectual currents in the nineteenth century were united. Whether we consider geology, zoology, political philosophy or the study of ancient civilizations, the nineteenth century was in every case the Century of History—a period marked by the growth of a new, dynamic world-picture.

This change took place simultaneously in a number of disciplines, scattered across the whole spectrum of science and scholarship, and it is unprofitable to ask which of them was the 'prime mover' in the change—which the 'horse' and which the 'carts'—for there was in fact a continuous interaction between them. In the early nineteenth century, educated men still kept abreast of important intellectual developments in all subjects, and a vigorous cross-fertilization of ideas was the natural consequence. While the central concepts of Darwinism stimulated the human sciences, Darwinian theory was itself built on foundations laid by the geologists, notably Lyell; and Lyell in turn acknowledged a stimulus from the new school of critical historians founded by Niebuhr at Berlin. So, having studied the historical revolutions within geology and zoology, we must now stand back and view them within a wider intellectual frame, relating them to the new historical consciousness which was affecting the whole of human thought.

Once again, the critical period was the twenty years between 1810 and 1830. Before this, men's overall vision of temporal development was still basically a 'teleological' one: the eighteenth-century philosophers had, in Carl Becker's striking phrase, 'demolished the Heavenly City of St. Augustine only to rebuild it with more up-to-date materials'. Everything that the mediaevals had attributed to divine Providence, they ascribed to Nature, while the moral demands of God's will were reinterpreted as deliverances of reason. But this was like putting an old picture in a new frame. The actual content of eighteenth-century thought had more in common with its mediaeval precursors than with our own post-evolutionary ideas. The Great Chain of Being was the sovereign hierarchy of Nature adapted to the theology of a new age, and representing God's 'chain of command' as the 'blueprint of creation' did not by itself alter men's beliefs within zoology. So, too, historians and philosophers around 1800 at first expressed their new awareness of temporal development in terms of inherited ideas, and failed to recognize the full complexity of the tale they had to tell.

In the years immediately before 1800, Johann Gottfried Herder was writing about historical development in the same manner as his contemporary, Lamarck, who was born in the same year. The course of cosmic history was, for both men, a single directed process which, as time went on, created more highly-developed organisms and societies in progressive sequence—so realizing (in Herder's words) 'the purposes of Mother Nature'. Herder thus took the first step towards a more historical conception of the world, but the second step was more difficult—that of recognizing that the actual course of history was too complex to be fitted into any simple teleological scheme.

Herder's ideas established a tradition of philosophical history which was developed during the next half-century by Hegel, Comte and Marx. Initially, this tradition kept the same teleological flavour. All the men concerned were convinced that the novelties of any particular era represented the fulfilment of some hidden purpose implicit throughout the earlier historical progression. The mainspring of this historical dynamic was located in different places by different philosophers. What Herder saw as the realization of the purposes of Mother Nature, Hegel interpreted as the self-development of *Geist* or Spirit—a reincarnation of Montesquieu's *esprit générale*, stripped of all assumptions about the fixity of human nature. Comte found the fundamental mechanism of social development in a cultural advance through which mythological habits of thought were successively displaced by philosophical and scientific ones;

while Marx looked rather to the changing economic relationships within society.

The resulting attitudes to history were a real advance on the static, eighteenth-century views. If so historically-minded a thinker as Herder never abandoned teleology, the heirs of the eighteenth-century Enlightenment did not even see history as a connected process at all. The Comte de Volney, for instance, was thirteen years Herder's junior; yet in his book, *Les Ruines*, he presented the successive civilizations of the world, not as a sequence of cultures, each of which built on the accumulated achievements of its predecessors, but merely as so many 'social machines', which had ceased to function because their citizens had not respected the eternal laws of human nature. For a man like Volney, the fundamental problems of politics were still concerned with social statics. The key-idea was not 'progression'—the successive appearance of novel relationships and functions: it was 'progress'—which, to the men of the Enlightenment, meant the deliberate establishment of a social equilibrium conforming to the demands of Reason. The resulting recipes for political reform were far too cut-and-dried: as though the conditions for universal happiness were specifiable in detail beforehand, and the consequences of a complete reconstruction of society could be foreseen and controlled.

Like all revolutionaries, the Enlightenment philosophers had dreamed of reforming society at a single stroke. Herder and his successors, on the other hand, understood that important social developments were frequently forced on mankind—by Lucretius' 'chance and necessity', Vico's 'wisdom of Providence', Kant's 'wiles of Nature' or Hegel's 'cunning of reason'. (In the 1960s, we are all freshly aware of this point, having witnessed the absolute antagonism between America and Russia unexpectedly generating the first tangible element of a new world-wide political system—the 'hot-line' telephone between Washington and Moscow.) They recognized also the limits of deliberate political change. Having a feeling for historical development and continuity, they could not envisage how any ideal political system could permanently fulfil all the potentialities of human nature, let alone how men could leap directly to it, by a deliberately-planned revolution. As they saw, 'revolutions' in political theory had some of the same defects as 'catastrophes' in natural history: the supposed changes were too drastic for the human mind to grasp or control, and the only revealing method was a uniformitarian one—to analyse the nature of social changes systematically and bring to light the inner continuities behind all political change. Effective social reform should be gradual, based on an understanding of these continuities and a recognition that 'human nature' is itself in course of historical development. (In this respect, the revolutionary enthusiasm of 'vulgar Marxism' is a backward step.)

Yet, despite their merits, the ideas of these philosophers are by now thoroughly dated. They did us a great service by insisting on the progressive character of temporal change, and the significance of those novelties which distinguished one epoch or era from another. But, after the establishment of modern historical criticism and Darwinian theory, it would be naive to suppose any longer that history represents either a *single* process, or one with a demonstrable *direction*. To regard the Ancient Hebrews, classical Greece and Rome, and Christian Europe, as the unique 'main-road' of history looks, from our point of view, too much like Lamarck's habit of selecting certain groups from each palaeontological epoch as 'the spearhead of evolution', and disregarding the lesser breeds. The course of history has been far more complex than that, and the continual interaction between different cultures precludes any possibility of identifying any 'march of time'. It is the same for the idea that history has a demonstrable direction: the most profound lesson of Darwin's work is that new creations of great functional significance often come into existence as by-products of processes, all of whose manifest goals lie in quite other directions; and the merits of these novelties depend, not on their conformity to any long-term historical tendency, but on their immediate appropriateness to the particular situation in hand. This is equally true for both organisms and institutions. If there is a key to the understanding of all history, it consists in recognizing not its single-directedness, but rather its multiple opportunism.

In this respect, even Marxism looked backwards to Lamarck rather than forwards to Darwin. It is not for nothing that Marx has been described as a 'Christian heretic': like Lamarck, he was in spirit the direct heir of the eighteenth-century Deists. In his system, the working-out of providential design kept its central place—only now it was rechristened the Dialectic. The random, opportunistic element in Darwinism was as repugnant to Marxists as it was to the more naive Christians: all partisans, whatever their ideology, find the sheer multiplicity of history an intellectual obstacle, which they are compelled to simplify rather than understand. By the 1820s, this ideological tendency among German philosophers had generated among German historians a reaction which was to create the new and more rigorous standards of criticism characteristic of modern historical scholarship.

Critical History and its Implications

This new interest in history affected men of all intellectual temperaments. The philosophically-minded hoped to recognize at once an overall pattern in temporal development; but others, with inclinations less speculative and more exact, felt that the more urgent task was to establish the actual facts about the historical past 'as they really were'. These words come from Leopold von Ranke's manifesto on the new method of critical history, published in 1824, in the preface to his *History of the Romance and German Peoples, 1494–1535*. Like the founders of the Geological Society of London, the critical historians saw that large-scale generalizations about the past must stand or fall by the evidence of the actual course of events. It was as premature to argue dogmatically about the causes governing the rise and fall of civilizations as it was to debate the unique claims of fire and water to be the universal causes of terrestrial change.

In fact, it was only just dawning on historians quite how untrustworthy their sources had hitherto been, and this feeling was intensified by the sense of period stemming from Vico and Herder. In his *Roman History*, for instance, Niebuhr argued that Livy's account of the early centuries of Rome was entirely anachronistic: it did little more than reproduce the patriotic legends current in the age of Augustine. One was accordingly forced to enquire how much original source-material for the early period—if any—Livy had actually had at his disposal. The lessons of this example were generalized by Theodor Mommsen for ancient history, and by Ranke for mediaeval and modern history. It was essential, first and foremost, to establish the true facts about any event by using only documents and reports contemporaneous with the events in question, and by asking oneself how much reliance could be placed on each of these 'primary sources'. Allowances must always be made for the interests, attitudes and prejudices of the men who composed the documents. Once these demands were pressed systematically, it became apparent just how little even the best eighteenth-century historical writing was rooted in the actual facts. Before long, indeed, professional historians realized that many of their most important primary sources were simply unavailable. Without access to State papers, which were still locked away in national archives or private libraries, most constitutional and diplomatic history could remain only so much guesswork, and it was not until well into the nineteenth century that either governments or ecclesiastical authorities recognized a general duty to admit outsiders to their records.

So long as national or ecclesiastical affairs have to be conducted confidentially, some conflict of interest between scholars and officials is unavoidable, and historians are still fighting today to reduce the delay in the publication of archives. Still, during the mid-nineteenth century historians were at last given a limited access to the State papers of the great powers, and by 1881 they were permitted to work even in the archives of the Papacy. With this opening of the archives, they were exposed to a flood of new documentary material, which has, if anything, grown in volume ever since. For more than fifty years, as a result, they were occupied almost exclusively with political and diplomatic history, and the critical scrutiny of past facts 'as they really were' became an end in itself. The original search for generalizations was forgotten, and the element of interpretation involved in any narrative statement of the facts was overlooked. It was hard enough—and ambition enough—to demonstrate conclusively what had really occurred. For the time being, Francis Bacon's hope that the study of history would be a prelude to the science of Man was deferred.

Meanwhile, the application of geological techniques to the study of human history was adding 'the testimony of things' to the evidence of documents and archives, so creating the beginnings of archaeology. This very soon exploded any simple, uni-directional theory of historical development. Scholars had always been aware in theory of the grandeur and antiquity of the ancient Egyptian civilization, but Napoleon's expedition to Egypt stimulated a new interest among French scholars in the early history of the country. The discovery of the 'Rosetta Stone'—a bilingual inscription, with parallel texts in Greek alphabetic writing and Egyptian hieroglyphics—gave Jean Champollion the means of deciphering the ancient Egyptian system of writing, and introduced scholars not merely to a brand-new literature but to historical archives of a centralized empire which was already old at a time when the Hebrews were still pastoral nomads. So Egypt and the Pharaohs ceased to be something in the margins of the Bible story, and acquired an historical significance in their own right. The same happened with the Assyrians, Babylonians and ancient Persians, through the work of Henry Rawlinson, Paul Botta and Henry Layard. The Persian King, Darius, had left a monumental inscription and bas-relief carved on a cliff-face at Behistun, three hundred feet above his main road from Mesopotamia to Central Asia. The text glorifying his memory was repeated in three languages, Persian, Elamite and Babylonian. With a mixture of physical courage and intellectual doggedness, Henry Rawlinson had himself lowered down the rock-face, copied the trilingual inscription, and used it to decipher the cuneiform script of ancient Mesopotamia. This feat became important soon afterwards, when Botta and Layard excavated the buried cities of ancient Assyria. For these excavations brought to light extensive historical archives written in this cuneiform script, and preserved in the library of King Ashur-bani-pal (seventh century B.C.).

At every point, the simple presuppositions with which scholars and public had approached the early history of Man were swept away. Layard's rediscovery of Nineveh caused a public sensation like that produced by the Dead Sea Scrolls in our own times, and he was acclaimed for the support his work apparently gave to the Biblical narrative. (He himself was gratified to confirm from actual inscriptions the authenticity of such kings as Tiglath-Peleser, whose name had been known only from an obscure reference in the Old Testament.) Assyrian art, too, for all its grandeur and magnificence, was at first regarded as a crude and primitive genre—a trial run for the true immortal glories of classical sculpture. But, as the material accumulated, it became more and more difficult to confine the Egyptians, Babylonians and Assyrians to the sidelines, or to patronize their artistic achievements. The 'main road' of history dissolved into a complex network, within which the Old Testament Hebrews formed only one, rather minor strand.

Since 1850, with the identification of at least half-a-dozen major cultures whose very existence had been entirely unsuspected, ancient history has become even more complex. The Sumerian civilization, including Abraham's own city of Ur, the Minoans of early Crete, the Hittites of Anatolia, the Indus Valley culture of Mohenjo-Daro—to say nothing about China: one after another, new elements uncovered by the archaeologist's spade have complicated our picture of Man's early history, and the beginnings of settled city-life. Just as the detailed analysis of actual strata circumvented the Biblical Flood and the argument between Neptunists and Vulcanists, so the detailed study of documents and sites has destroyed the assumptions on which earlier interpretations of human history rested. By the late-nineteenth century, too much was known about the actual facts to support any longer a belief that human history was the unique expression, either of Providence or of Progress.

The Evolution of Humanity

This revolution within human history owed very little to Darwin, for by 1859 much of it was already over. Niebuhr moved to the newly-founded University of Berlin in 1810; Ranke belonged to Lyell's generation rather than Darwin's; Rawlinson and Layard did their most important work in the 1830s and '40s. From this point of view, indeed, the *Origin of Species* was a late phase in a more extended intellectual operation, which brought the new historical categories to bear on one particularly awkward case.

Nor were the first attempts to apply Darwinian ideas to moral and political affairs very impressive. In social and political theory, the authority of Darwin was invoked only to lend support to positions already adopted for other reasons. There were men who picked on the violence and conflict of the 'struggle for life' as a justification for militarism. War was a legitimate instrument of social advance, by which higher varieties of the human species ('super-men') would establish themselves at the expense of their biological inferiors. Without going this far, there were other men who saw in Darwinism the justification for unrestrained social and economic competition, as a means to the long-term betterment of the human race. In all such arguments, however, the assumed analogy between organic evolution and social development was insufficient: this was not 'Social Darwinism' but rather, as D. C. Somervell has called it, 'pseudo-Darwinism'. Looking back on this period in later life, T. H. Huxley emphasized that the requirements of 'biological advance' were in certain respects opposed to those of 'social progress'. Between the total population of living creatures and the social order of the human species, there was, in fact, very little analogy; so there was nothing in the sphere of morals and politics corresponding to the action of 'natural selection'.

It was the sheer diversity and multiplicity revealed by history and archaeology that gave new life to the human sciences, and stimulated a fresh critical approach to accepted beliefs and institutions. Like the first Ionian philosophers, the men of nineteenth-century Europe were compelled to admit that their own traditions were far from unique, since civilizations had existed whose customs and institutions were quite different from their own. Even the central documents of their religion had the failings of other human documents: the combination of comparative religion and the 'higher criticism' of Biblical texts was a powerful solvent of inherited assumptions. This recognition of the diversity of belief, institutions and customs did not by itself teach men how to live or order their affairs. But, by removing opposition based on sheer conservatism alone, it did much to pave the way for functional arguments and utilitarian reforms. Institutions could no longer be justified by appeal to the wisdom of the ages: they must have some demonstrable social value.

Yet, when all is said and done, the theory of evolution did give a real stimulus to the human sciences. Above all, Darwinism finally broke the bounds within which natural scientists had worked since the seventeenth century. When, for instance, Descartes contrasted the 'mechanical' properties of Matter with the 'rational' character of Mind, he removed Man's rational nature from the scope of the World-Machine. Similarly, when Leibniz declared that the world of Nature was entirely self-sufficient, he did so in order to emphasize the essential contrast between Nature and Man. This compromise was the price that, from 1650 to 1850, physical scientists consented to pay in return for the right to develop a mechanical picture of Nature. For all but a few heretics, such as de la Mettrie, Man was a 'singularity' in Nature—the point at which the general laws of mechanism ceased to tell the whole story. It is true that social and economic theorists had spoken of society as maintained by a system of 'social forces', analogous to those of the physical world, but this had never been more than an analogy. The two worlds of Man and Nature, of Reason and Mechanism, were still treated as parallel and distinct realms.

Yet, after Darwin, how much longer could one retain an absolute distinction between rational Man and mechanical Nature? Though in writing the *Origin* Darwin limited his discussion to other kinds of living creature, the implications of his theory for the human species were as plain as a pikestaff. One could scarcely use variation and natural selection to explain the production of all other plants and animals, only to make an exception in the case of Man; and this Darwin conceded in a passing remark—

> In the future I see open fields for far more important researches . . . much light will be thrown on the origin of man and his history.

Once man himself was accepted as a natural product of the evolutionary process, the rest of the Cartesian compromise could hardly be maintained. It was this obvious extension of the Darwinian theory, rather than the actual argument of the *Origin*, which was the occasion for Bishop Wilberforce's scurrilous attack at the British Association meeting of 1860. Men were, in fact, on the verge of discovering the antiquity of their own species, the evolutionary genealogy of their forefathers, and the prehistoric epochs through which they had developed.

So, while on the public stage the religious debate about evolution rumbled along, scientists behind the scenes were collecting the first materials for human palaeontology. Despite the work of the archaeologists, the Biblical chronology had hung on until the 1850s as a measure of the earthly existence of Man. The complete absence of human relics from all but the uppermost geological deposits made it possible to argue—in line with Eus bius' time-scale—that Man, at any rate, had been created around 4000 B.C.: the first evidence that the human species had a much greater antiquity was quietly overlooked. Orthodox scientists were unwilling to accept as authentic the discovery in eastern England in 1797 of flint tools and weapons associated with fossil bones of large extinct animals: these were so out of harmony with the other findings of palaeontology that, for the time being, everyone agreed to ignore them. Only in the 1850s did the evidence become unanswerable: human beings had demonstrably existed alongside the extinct vertebrates of earlier epochs.

The man who compelled official science to change its mind was Boucher de Perthes, a customs official from Abbeville. Having recovered large numbers of worked flints and bones from the gravel beds of the Somme, he struggled for more than twenty years against the scepticism of the learned world: the authenticity of his finds was accepted only when two parties of English scientists, including Charles Lyell himself, visited Abbeville to check his results. It was, indeed, with Boucher de Perthes as much as Darwin in mind that Lyell, now in his sixties, embarked on his second major contribution to science: *The Antiquity of Man*, which appeared in 1864. Darwin and Huxley had been bitterly disappointed that the great man had not hitherto embraced the theory of evolution publicly, but the *Antiquity* did something to remedy the situation. All along, Lyell's main reason for rejecting organic progression—to say nothing of evolution—had been the fragmentary and ambiguous character of the fossil evidence: this lack of continuity had been particularly evident in the case of the human species. Now he reversed his position, and to Darwin's irritation he gave much of the credit to Lamarck, whose arguments had been refuted in all earlier editions of the *Principles of Geology*. Now that he was finally convinced that the palaeontological status of man was the same as that of any other animal species, Lyell at last dispensed with all 'special creations', and revised the subsequent editions of his *Principles* accordingly.

Meanwhile, fresh discoveries were bridging the gap between man and the lower animals. The first bones to be accepted as belonging to a transitional form were found in Germany in 1856, in a cave near the River Neander. With its receding forehead and prominent brow-ridges, the skull of this 'Neanderthal' man had a decidedly ape-like appearance: so much so, that the sceptical dismissed it as that of 'a pathological idiot', or of a 'rickety Mongolian Cossack'. But, by 1870, similar remains had been recovered from other parts of Europe; also skeletons intermediate between the Neanderthal and modern forms, beginning with the Cro-Magnon man from Les Eyzies in the valley of the Vézère inland from Bordeaux. So Darwin embarked on his *Descent of Man* (1871) in the knowledge that, among his scientific colleagues at least, the worst of the storms had passed and he was free to discuss the evolutionary origins of the human species in a dispassionate and scientific manner.

On the physiological and zoological planes, man now took his place alongside the other animals. But to concede that the human species had an evolutionary ancestry relating it back to earlier, non-human forms, raised inescapable problems for psychology and social theory also. Even the earliest archaeological finds already showed men living in highly-organized states, with elaborate social structures and legal systems. Writing itself, without which no historical documents could exist, appeared to be a comparatively late development. Now, however, men were forced to embark on enquiries into the development of human behaviour, both individual and social. For the first time, there was a serious hope that the speculations of Diodoros and Lucretius—about the gradual development of human society, from cave-dwelling to cities—might be given a proper scientific foundation. Under the leadership of Sir John Lubbock (later Lord Avebury) whose book on *Pre-Historic Times* appeared in the year 1865, archaeology set out to reconstruct not merely the skeletons of earlier men but also their ways of life.

A century of work in these new directions has done much to fill in the gaps in our historical perspective: notably, to break down the earlier discontinuities separating the temporal development of the natural world (which was the concern of cosmology, geology and zoology) from human history (which interpreted documents). The evidence all confirmed Vico's picture of a gradual, painful ascent out of a primitive state into comparative civilization. As a result, Montesquieu's static picture of social structure finally collapsed: existing societies were not the manifestation of a constant 'human nature', but the end-products of a sequence of historical phases, by which the life of anthropoid apes and 'cave-men' had been slowly transformed into the more complex life of modern industrial society. So vertebrate and human palaeontology combined to tell a single story of continuous development which, in its most recent phase, merges into pre-history, classical archaeology and, finally, orthodox documentary history.

This continuous perspective is accompanied by a consistent chronology. The antiquity of the human race, or at any rate its tool-using precursors, is now believed to cover the last million years. From 20,000 years back, the stone tools are carefully worked and of high quality. The first metals were already being worked at least five thousand years ago: copper and gold originally but, beginning among the Hittites around the fourteenth century B.C., iron also. A great deal remains to be discovered about the origins of city life, the development of early culture and technology, and so on; but, in outline, the chronology of historical development, as revealed by prehistory, archaeology and the other human sciences, appears to be well established.

At two significant points, however, obscurities remain. If we look back at human history proper, we may feel that the accumulation of authenticated facts has not yet been matched by adequate theories. The process of establishing the facts about the past is still going on but, when it comes to interpreting these facts in terms of ideas drawn from the human sciences, comparatively little has been done. Economic history represents the most serious attempt at such an interpretation; but the relations which history should bear to anthropology, sociology and the rest are still not entirely clear. This situation is probably largely due to the continued weakness of the human sciences themselves: it is as though geology had been forced to remain purely descriptive, for lack of any adequate support from physical and chemical theory. Yet it seems to be due in part also to an uncertainty among historians about the *kinds* of explanation which they can legitimately give. Must their aim be solely narrative? Or can they 'account for' events, by relating them to general principles? And if they do attempt to 'explain' events and institutions, should they do so by relating them *functionally* to other aspects of their own times, or by relating them *temporally* to earlier events and institutions of which they are the outcome? How far there are 'invariants' in social and economic history, and how far all institutions are in flux, historians still do not agree.

At this point, indeed, a better analysis of the problems at issue is probably required, if history is ever to become an 'explanatory' science. Any direct opposition between functional explanation (in terms of contemporary factors alone) and dialectical explanation (in terms of a temporal succession) would certainly be an oversimplification. As Darwin's analysis made clear, the establishment of novel variant forms is intelligible only if one takes into account *both* the succession of events by which they come into existence *and* their adaptedness to the environments in which they actually survive. And what is true of organisms can, with appropriate safeguards, be said of social institutions also: we shall understand the nature of any institution only if we can explain *both* how it is related back to earlier social forms *and* what social role it played in its own time.

In psychology, again, the intellectual revolution which culminated in Darwin's theory still has some way to go. Eighteenth-century philosophers contrasted the social habits of human beings with the mere gregariousness of animals—treating the latter as the product of instinct, yet at the same time hailing the former as the triumph of reason. So long as the uniqueness of man remained unquestioned, this contrast could stand unchallenged, but in the aftermath of evolution-theory any absolute opposition between human reason and animal instinct begins to appear a decided exaggeration. The crucial point is not that human behaviour is less rational than our forefathers believed: it is, on the contrary, that many animals already display in embryo form those modes of behaviour which are quoted to prove the rationality of Man. But animal behaviour has been studied in systematic and scientific ways only in the last few years, and the science of 'ethology' is still in its infancy. The full lessons to be learned from comparing, instead of contrasting, the behaviour of humans and other animals remain a matter for the future.

Pressed to its final conclusion—that human behaviour has developed as smoothly and continuously from that of our evolutionary ancestors as human anatomy and physiology—the Darwinian integration of Man into Nature still encounters resistance even today. In the psychological field, many people still believe that Man, in virtue of unique mental and moral capacities, must be separated from the rest of Nature. There, at any rate, Pope Pius XII drew the line quite explicitly in the encyclical, *Humani Generis* (1951):

> In conformity with the present state of science and theology, the doctrine of evolution should be examined and discussed by experts in

both fields, in so far as it deals with research on the origin of the human *body*, which it states to come from pre-existent organic matter: the Catholic faith obliges us to believe that *souls* were created directly by God.

The Flux of Nature

In the course of the nineteenth century, then, natural history—in its original sense of the history of Nature—was integrated on the one side with human history, and on the other side with natural science proper. Before 1750, cosmology, zoology and geology had been three distinct and independent realms of thought, having few problems in common either with each other or with the study of human affairs. By 1900, they were all concerned with temporal processes of familiar kinds taking place over longer or shorter periods. It was now clear that the 'testimony of things' could be as reliable a guide to the past as human testimony. Beyond a certain point, indeed, the evidence of Nature was infinitely more reliable than memory or legend: properly interpreted, it provided a secure way of arguing back far beyond the time-barrier of traditional thought. This, however, was true only on one condition, which applied with equal force to all historical sciences—human and natural alike. We must treat the past as continuous with the present, and interpret the traces left by earlier events in terms of the same laws and principles as apply in the present era.

That was the wisdom of the uniformitarian *method*. Yet, as time went on, the actual history to which all these enquiries led became less and less uniform in fact. All the constant features of our everyday human world—social institutions, organic species, the Earth's crust, and the planetary system itself—turned out to be in continuous slow flux. The problem was to reconcile theoretically-uniform causes with historically-variable effects. In 1790, James Hutton could still dismiss the cosmological speculations of Leibniz and Buffon as irrelevant to geological theory, and insist (like Aristotle) that geology revealed 'no Vestige of a Beginning': eighty years later, however, no scientist doubted that geology and cosmology must come to terms, and agree on a consistent account of the Earth's development. Likewise in astronomy: Newton's conviction—inherited ultimately from Plato—that the Order in the Solar System had been placed there in the original Creation woke no echo among nineteenth-century astronomers, who were universally convinced that the composition and movements of the planets must eventually find some physical explanation. Even less could nineteenth-century scientists accept the Stoic picture of the Earth and the planetary system as subject to a recurrent cycle of creations and destructions. The method of thought which proved its worth in science was that of the Ionians and Epicureans, for whom cosmic development was a continuous and continuing process, still capable of producing genuine novelties—a philosophy of Nature against which well-meaning European thinkers had struggled for centuries, under the impression that it was inevitably 'materialist' and 'atheistical'.

At their widest, the repercussions of the historical revolution influenced men's attitudes to the very remote past in two ways. In the first place, it ceased to be merely *hypothetical*, an object which perpetually eluded the grasp of rational thought. In earlier times men could think about the past only in metaphysical terms—by arguing *a priori* about what 'must-have-been'—or in theological terms, by basing their conclusions on Revelation ('And God said: . . .'). Now they began to establish in a piecemeal, scientific way, what course historical development had in fact taken; and the results departed from all the products of dogma and speculation to an unexpected extent. In the second place, the time-dimension itself now acquired a significance more universal than it had for the early Christians. In marrying the Jewish view of human history, as the working-out of a divine covenant, to ideas about Nature derived from the Greek philosophers, they had seen time as significant only for humanity. Nature had no history: it was merely the stage on which the drama of the Redemption was enacted. Now, historical development—or *evolution*, in its wider sense—acquired an importance for both the natural and the human sciences which previously it had only for theology. Everywhere, as a result, the *a priori* patterns of thought found in Plato, Aristotle and the Stoics began to be replaced by more empirical, developmental theories. Society had neither existed for ever in its present form, nor been created suddenly by a single act; nor, for that matter, had there been eternal cycles of civilization and decay, as suggested by Plato's *Critias* and some Renaissance humanists. Human life and society have taken on their present form gradually, and the archaeologist and historian must unravel the successive phases of this process.

The development of Nature, too, was now regarded as a continuing one-way process. Organic species have neither existed for all Eternity, nor were they created in their present form at a stroke; nor have they been caught up in a perpetual cycle of creations and destructions. On the contrary, they too are the current end-products of a multi-million-year evolution, in which earlier and more primitive organisms have gradually been superseded by later, better-adapted creatures. Similar conclusions apply to the history of the Earth. In none of these cases has our understanding been improved by asserting, *either* that the systems in question have existed for ever in a Steady State, *or* that they were created out of nothing in a Big Bang at some moment of the remote past, *or* that they conformed to a Cosmic Cycle of creations and destructions. In each case, men grasped the structure of the past only when they took a genuinely historical view; which meant interpreting the present states of Nature and humanity as temporary products of a continuing process developing through time.

FURTHER READING AND REFERENCES

There are exhaustive accounts of the history of historical writing by J. W. Thompson, H. E. Barnes and E. Fueter. Classic discussions of historical thought in the nineteenth century, from three different points of view, are to be found in

G. P. Gooch: *History and Historiography in the 19th Century*
R. G. Collingwood: *The Idea of History*
F. Meinecke: *Die Entstehung des Historismus*

See also the more philosophical discussions in

W. Dray: *Laws and Explanation in History*
J. Ortega y Gasset: *History as a System*

The rise of archaeology is discussed in *A Hundred Years of Archaeology* by Grahame Clarke, and an entertaining sidelight on this subject is Gordon Waterfield's biography, *Layard of Nineveh.*

See also the anthology

Man's Discovery of his Past: ed. R. F. Heizer, which records an early use of the word 'prehistoric' by Tournal (1833), anticipating the general rise of archaeology by twenty years.

For the general nineteenth-century development, see

D. C. Somervell: *English Thought in the 19th Century*
T. H. Huxley: *Man's Place in Nature*
W. Irvine: *Apes, Angels and Victorians*
John C. Greene: *Darwin and the Modern World-View.*

II

Time and the Physical World

BY THE end of the nineteenth century this historical transformation had penetrated deeply into all fields of natural science—except one. In the geological, biological and human sciences the *a priori* patterns of Greek philosophy had everywhere been displaced, but the new developmental categories influenced physics and chemistry much more slowly; and now, more than sixty years later, it is still an open question whether physics will ever become a completely historical science. The physical sciences had stood aside from the historical revolution which transformed the rest of natural science, taking it as axiomatic that certain aspects of the world remained fixed and permanent throughout all other natural changes; and though, by the mid-twentieth century, the list of these timeless entities—or 'eternal principles', as the Greeks had called them—is much shorter than it was in 1700, the existence of unchanging physical laws, at least, is still regarded as one enduring aspect of the natural world.

During the eighteenth century, the orthodox picture of physical Nature was that stated by Isaac Newton at the end of the *Opticks*. This involved permanent features of five different kinds. First, there were the 'solid, massy, hard, impenetrable, moveable Particles' that constituted the ultimate physical population of the natural world. It was inconceivable that these particles should change—at any rate until God resolved to terminate the present Order of Nature entirely. One of the crucial properties of these particles was their 'force of inertia'—what we should now call their 'mass'; and the second timeless feature of the world comprised those 'passive Laws of Motion as naturally arise from that Force [of inertia]'. By this Newton meant the laws expressed in the axioms of his dynamics—and again, within the present dispensation, these laws of motion would not vary from age to age. Thirdly, Newton drew attention to 'certain active Principles' associated with gravity, magnetism, electric attraction and 'fermentation'—i.e. chemical reactions. These principles, too, conformed to 'general Laws of Nature' determined by God in the original Creation. For example, the 'inverse-square law' of gravitation was part of God's timeless design for Nature, and neither the power nor the constant in the formula would vary without His direct intervention. The fourth permanent feature in the Order of Nature was the comparative stability of the planetary orbits; the fifth and last was the adaptive pattern in organic structure—what he called 'the Uniformity in the Bodies of Animals'. There was, Newton taught, no logical necessity about this particular picture. God was free to fix the laws of Nature just as He pleased, and it was quite conceivable that, in different parts of the cosmos, He had fashioned systems working according to other laws. But, so far as our own region was concerned, the Order of Nature had evidently been created on a fixed pattern of indivisible particles, laws of motion, active principles, planetary orbits and organic uniformities.

Only in one respect was the Newtonian framework eroded during the eighteenth century. By the 1780s Kant's speculation that, on the astronomical scale, the creation of the natural order might still be in progress, seemed to be finding some real confirmation. William Herschel, who had built the most powerful telescopes yet available—with which he discovered the planet Uranus in the year 1781—soon satisfied himself that many of the so-called 'fixed' stars were in fact in motion. Presumably gravitation acted not only within the Solar System, but between all members of the astronomical universe, and one result of this gravitational interaction could well be the concentration of stars into clusters and galaxies. With the help of a striking series of observations, he sketched out a possible account of the processes by which stars and star-clusters might evolve—the stars being formed by the agglomeration of faintly luminous inter-stellar dust and gas (which he first observed in 1790), then being drawn together into more and more compact associations, some of which eventually collapsed together in a final blaze of glory. This theory was immediately applied by Laplace to explain the origin and properties of the Solar System. That, too, could have formed by condensation out of an original diffuse cloud of matter, and its comparative stability required no other, supernatural origin. So, by 1800, one item had been removed from Newton's list of divinely-ordained principles and uniformities.

In other respects, however, the list remained unchanged. If anything, it became even more firmly established during the years that followed. For the 'permanent Particles' of Newton's physical world were now identified with the 'atoms' and 'molecules' of nineteenth-century physics and it was widely accepted—at any rate among physical scientists—that God had originally created particles of some ninety different fixed kinds, which would persist unchanged throughout all the transformations of the physical world. In his address to the British Association in 1873, James Clerk Maxwell made the theological affiliations of this doctrine quite explicit:

> No theory of evolution can be formed to account for the similarity of molecules, for evolution necessarily implies continuous change, and the molecule is incapable of growth or decay, of generation or destruction. None of the processes of Nature, since the time when Nature began, have produced the slightest difference in the properties of any molecule. We are therefore unable to ascribe either the existence of the molecules or the identity of their properties to the operation of any of the causes which we call natural. . . .
>
> Natural causes, as we know, are at work, which tend to modify, if they do not at length destroy, all the arrangements and dimensions of the earth and the whole solar system. But though in the course of ages catastrophes have occurred and may yet occur in the heavens, though ancient systems may be dissolved and new systems evolved out of their ruins, the molecules out of which these systems are built—the foundation stones of the material universe—remain unbroken and unworn.
>
> They continue this day as they were created—perfect in number and measure and weight, and from the ineffaceable characters impressed on them we may learn that those aspirations after accuracy in measurement, truth in statement, and justice in action, which we reckon among our noblest attributes as men, are ours because they are essential constituents of the image of Him who in the beginning created, not only the heaven and the earth, but the materials of which heaven and earth consist.

The essential framework of the nineteenth-century physical universe thus consisted of fixed atoms, interacting by forces which conformed to fixed laws. Everything else was derivative and transient; and it is ironical that the effect of Dalton's and Maxwell's theories of matter was to freeze the central categories of physics and chemistry into a rigidly a-historical form at the very time that biology was being transformed into a mature, historical science by the publication of Darwin's *Origin of Species*.

During the twentieth century, the list of changeless physical entities has drastically shortened. The discovery of radioactivity raised the first doubts about the permanence of the atom, and this was only the first pebble initiating an intellectual avalanche (see chapters 12 and 13 of *The Architecture of Matter*). Though the physical world, as envisaged in modern quantum theory, is still composed of so-called 'fundamental particles', these represent a much more transient and changeable population than either Newton or Maxwell could ever have contemplated. All of them, indeed, are transformed in appropriate circumstances into others, and by now there remain no 'eternal units' of matter or energy at all. Some of the physicists' particles may seem more fundamental than others; but it begins to appear that their inter-relations will be best understood when they are seen as by-products—temporary configurations in the fields of force which are defined by the mathematical equations of physical theory.

Newton's original five categories have thus been cut down to one: the fixed laws of Nature. It is these whose permutations—in theory—govern the whole kaleidoscope of matter and energy which is the scientist's object of study. Given an appropriate and fixed system of equations, one can hope to explain physical processes from the sub-atomic up to the molecular, and macroscopic, and even the astronomical, scale; and this can be done without exempting any physical objects whatever from the flux of time—which means accepting that there *are* no permanent material 'atoms', in the original sense. For contemporary physicists, indeed, the particles making up ordinary matter are so many temporary consequences of the equations governing the fundamental force-fields, just as for Laplace the structure of the Solar System was a consequence of universal gravitation. The outstanding question now is, whether the laws of Nature themselves—the last *a-historical* feature of the physicists' world-picture—will in their turn prove to be subject to the flux of time.

The Evolution of Stars and Chemical Elements

The only region of science in which this question can seriously be raised at the present time is physical cosmology, for there we are concerned with processes taking place over thousands of millions of years, during which any slow alteration in the fundamental laws of Nature might become evident. In any case, the character of cosmological theory has at all stages been affected by changes in the central categories of contemporary physical theory; and the encroachment of evolutionary ideas into physical thought during the twentieth century shows up clearly in this, the most 'historical' of the physical sciences. If we follow out the development of cosmological ideas over the last 150 years, we can watch this historical consciousness permeating the science, and examine the nature of the evidence which may eventually remove the last timeless features from our conception of the physical world.

Laplace declared that a sufficiently powerful mind, if given the initial positions and velocities of all the atoms in the universe, could in principle calculate all its subsequent history. The life-story of the material world was thus a matter for dynamics alone, and during much of the nineteenth century physical scientists continued to regard it in this light. The revolution in chemical thought initiated by Dalton and Berzelius was slow to leave its mark on cosmological speculation. Chemistry was primarily a laboratory science, concerned with material substances which we can actually handle. Meteorites alone provided direct chemical evidence of the properties of 'celestial matter', and it seemed that the chemical constitution of the stars would for ever elude the scientists' grasp. This particular obstacle was overcome by the development of the spectroscope: it was found that the light-spectrum of every star contained a characteristic pattern of lines which corresponded precisely to lines already familiar in the spectra of terrestrial substances. But, although the light-spectra of different heavenly bodies could reveal the chemical substances present in them, their life-histories remained as mysterious to nineteenth-century astronomers as they had been to Lucretius.

The essential clues to the processes of star-formation and development turned out to be, not chemical, but sub-atomic. Spectrum analysis had to wait until the twentieth century for its rationale, when Bohr explained atomic spectra in terms of 'quantum jumps' within the atom, and the chief source of energy in the stars is now believed to be the processes of 'thermonuclear fusion' (see once again *The Architecture of Matter*). As a result of this convergence of nuclear physics and astronomy, the evolution of stars and galaxies has, since 1945, become a serious and fruitful field of scientific enquiry. Like Linnaeus in zoology, earlier astronomers had built up a comprehensive taxonomic scheme, in which stars, star-clusters and galaxies were classified according to their observed shapes and light-spectra; but now one can begin to reinterpret this 'Linnaean' classification—as Darwin did for plants and animals—in terms of a temporal sequence, the development of stars involving a standard succession of nuclear reactions, and their taxonomic status being determined by the stage which they have reached.

The analogy implied by the use of the word 'evolution' is still only partial. Firstly: so far as we know, there is nothing in astronomy corresponding to 'natural selection' which could lead to the preferential survival of any particular type of celestial body. Secondly: there is nothing in cosmic history up to now which would correspond to the replacement of one species by another within a population. Thirdly: the theory of 'stellar evolution' at present deals only with the characteristic life-cycle of individual stars and star-clusters, not with a universal 'cosmic history' embracing the whole course of astronomical affairs. If we wish to form a proper judgement of twentieth-century cosmological theory, in fact, we must keep these two aspects separate, looking first at the evidence for stellar evolution, and then seeing how cosmologists have attempted to extend our understanding to the origin of the universe-as-a-whole.

The arguments for stellar evolution begin from astronomical observations of three general types. For a start, astronomers have found in the sky a marked contrast which, since the inauguration of the two-hundred-inch telescope on Mount Palomar, has become quite unmistakable. On the one hand, there are regions characterized by diffuse and nebulous clouds, some luminous, others visible only as dark shadows against a luminous background; all these consist apparently of interstellar dust and gas. Much of the region surrounding our own Sun appears to be like this, and so also are the 'arms' of the spiral galaxies. Within such regions the most luminous bodies are the so-called 'blue giants', whose surface temperature is so high that they appear, not just red or white, but positively *blue*-hot; and the spectra of stars in these regions show evidence of a high metal content. On the other hand, there are regions of sky apparently devoid of inter-stellar gas, and populated by celestial objects of quite a different kind. These regions include the elliptical galaxies, and the central cores of the spiral galaxies. (It has, in fact, been suggested that certain elliptical galaxies may be former spiral galaxies which have lost their arms.) Within these regions, the typical stars are poor in metals, and the most intensely luminous bodies—the 'red giants'—are much cooler than those in the other regions.

Secondly, within the 'gas-filled' regions, one can find a great variety of bodies, ranging from the most highly diffuse clouds to fuzzy objects almost as brilliant as normal stars. These more or less diffuse objects are accompanied by genuine stars having an enormous range of colours, and so of surface temperatures. If we compare the luminosity and the colour of all these different bodies, we find nearly all of them conforming to a common pattern: the fainter, the redder. A few exceptions apart, the stars with the highest surface temperatures also have the greatest intrinsic brightness.

The third important observation has to do with the few exceptions just mentioned. If we study a number of localized star-clusters (for instance, the Pleiades, the Hyades and the Hair of Berenice) and consider how the relationship between luminosity and colour applies to them, we find that the general pattern in each case is precisely the same: the differences between the clusters show up only in the exceptions. In any particular cluster, there are a few comparatively intense stars, whose light is redder than one would expect for their brightness—a few isolated 'red giants', in a region where 'blue giants' are the general rule. In each cluster, these exceptions form a distinct sequence, having their own particular relationship between luminosity and colour, which is specific for that cluster.

The accepted interpretation of these observations is based on the assumption that the raw material out of which stars are formed consists of inter-stellar dust and gas, predominantly hydrogen gas. Star-formation is actively going on at the present time in the arms of spiral nebulae and other similar regions: the blue-hot stars characteristic of these regions are the newest, and so the youngest stars. Their hydrogen has just condensed under gravitation to the point of 'igniting', and is 'burning' into helium with its pristine vigour. (The fuzzier areas—some luminous, some opaque—possibly represent condensations of hydrogen which are just forming, or which are in process of igniting.) Generally, stars are not born out of inter-stellar gas one at a time: instead, a local concentration of hydrogen condenses into a compact cluster of stars, all of which are—give or take a million years—much of an age. Having ignited, these stars all begin to run the same general course. During this phase, their hydrogen is converted steadily into helium, and this process continues until about one-tenth of the mass has been transformed, though the rate at which the conversion actually takes place depends on the initial mass of the star. Those with the largest initial quantity of hydrogen burn very much faster.

Eventually, the accumulation of helium at the core of the star produces an instability. Two things now happen: the helium core contracts, and at the same time the surrounding envelope of hydrogen balloons out to many times its previous size. As a result, the outer shell cools, even though the hydrogen is still being converted at a great rate. The star now appears to us unusually red for its luminosity—it forms, in fact, a 'red giant'. Meanwhile, the helium core collapses under gravitational attraction and its temperature rises to a point at which fresh nuclear fusion is possible. The helium nuclei now combine into such elements as carbon and oxygen, and these in turn interact with the helium itself, fusing again to form other nuclei, and so other chemical elements, up to and including iron (atomic weight, 56). The direct formation of elements heavier than iron is no longer possible by the fusion of helium, but enough stray neutrons will be present in the core to transform the iron nuclei, step by step, into the whole gamut of more complex elements. So, astro-physicists now believe, the central cores of 'red giants' serve as factories for the production of heavier chemical elements.

In terms of this theory, one can now estimate the age of any particular star-cluster, such as the Pleiades. In a cluster which has formed only very recently, even the largest stars will not yet have run through their initial phase, and 'red giants' should be completely absent: this is the case with the cluster NGC 2362, which appears to be two million years old at most, and may still be in process of forming. Where only the very biggest stars in a cluster have reddened, the cluster will be a mere 100 million years or so in age. The older the cluster, the greater the proportion of stars which will have passed the critical point and begun to turn into 'red giants': one can estimate the age of the cluster by seeing just how far this reddening tendency has progressed. On this basis, it is calculated that the Hyades condensed out some 400 million years ago, the Hair of Berenice 300 million, and the Pleiades about 20 million years ago.

The subsequent life-history of stars formed in this way is less clear, but is understood in rough terms. For a time, the star may pulsate, but eventually the nuclear matter in the core collapses together, and is compressed to the ultimate limit in a complete physical fusion. This apparently happens, however, only to the smaller stars. Those which are too big explode, throw off their surplus matter in the temporary bursts of luminosity we call 'novae', and leave behind the last cooling embers of the defunct star—a 'white dwarf'. Alternatively, the final explosion may be even vaster, forming a 'super-nova'. Then, not even a white dwarf remains: the whole substance of the star is scattered into the surrounding space as non-luminous dust—as seems to have happened to the super-nova which appeared in the constellation of the Crab in A.D. 1054. So the cycle of star-formation is free to begin again.

The situation in the elliptical galaxies is very different. There, it seems, all dust and gas have somehow been swept away, and the only stars appear to be old and comparatively small. The process of star-formation having long since ended, there are no bright young blue-giants, and the surviving members of the galaxy calmly go through the last stages of a star's life-cycle; though, from time to time, as they finally reach the point of instability, a few of them flare up as red-giants. Likewise, the central cores of the spiral galaxies are composed largely of aged, reddish stars, which contrast with the younger, 'bluer' population of the dust-filled arms. Some of the outer dust may, indeed, have come from the explosion of older stars in the core. The ejected matter contains a great variety of chemical elements—notably, the metals—and these are available to enter into the composition of the younger stars in the spiral arms. So the current theory succeeds in accounting for the final contrast between the two main populations of stars: the younger stars in the arms of the spiral galaxies, whose spectra reveal fair quantities of metals and other complex elements, and the older stars of the central regions and the elliptical galaxies, composed predominantly of hydrogen and helium.

This interpretation—which is worth comparing with Kant's theory (see Chapter 6)—ties together most elegantly a remarkable range of astronomical facts, and introduces a genuine time-dimension into astronomy. It is an account which, at one end, connects naturally with theories about the formation of the planets—whose matter was probably captured by the Sun from surrounding inter-stellar dust; at the other end, it enables us to extend our time-scale back for at least 7000 million years, to the first formation of the Sun and the oldest of the neighbouring star-clusters. But, beyond this point, the story of stellar evolution comes up against the problem of the origin of the entire cosmos.

Astronomy and the Problem of Creation

We must, accordingly, turn at this point to the lines of argument by which astronomers and physicists have, in recent years, attempted to establish the history of the cosmos. These are more concerned with questions about the first origin—if any—of the universe-as-a-whole than with the temporal development of particular celestial bodies. One warning is necessary: over these cosmological theories there is nothing like the same kind of agreement among astronomers that exists over theories of stellar evolution. In cosmology, the chief protagonists take up standpoints which are sharply opposed, and even—as at present formulated—irreconcilable. Some see the astronomical evidence as supporting a belief that the entire universe began at an initial moment in time about 10,000 million years ago, through a cataclysmic Creation, by which time and matter came into existence together, once and for all. Others believe that the cosmos has had an unlimited existence in time, and its average state and appearance have always been similar to what they are today. This uniformity-throughout-eternity is even presented at times as a necessary axiom of all scientific thought about cosmology—a rational presupposition, to which any acceptable account of the universe must conform. A third party has adopted yet another point of view. The cosmos has neither had an initial Creation, nor displayed an eternal changelessness: instead, it has passed through a recurring cycle of similar changes, oscillating between two extremes, with an overall period of perhaps 100,000 million years.

The disagreement between supporters of these views today is just about complete. Nor does there seem to be any real hope of reaching an accommodation without abandoning elements which are regarded as indispensable to the theories. This theoretical impasse is as obstinate as those which afflicted geology around 1800 and zoology in the 1830s and '40s, and we must ask in due course whether anything can be learnt from the manner in which those other sciences achieved a genuinely historical standpoint.

Each of the three rival accounts gives its own interpretation of the crucial piece of astronomical evidence—namely, the 'red-shift' in the light-spectra from distant celestial objects. If scrutinized with a good-quality spectroscope, these spectra display lines and bands spread across the gamut of wave-lengths, in patterns characteristic of different kinds of atoms and molecules (Plate 10). But there is one striking difference between the spectrum we obtain (say) by exciting hydrogen in a laboratory, and the corresponding spectrum as observed in the light from a distant galaxy. In the astronomical case, the whole pattern of spectrum-lines is shifted towards the longer wave-lengths—that is, towards the red end of the visible spectrum. Moreover, there appears to be a close relationship between the magnitude of this red-shift and the distance of the galaxy in question: the farther away the galaxy, the more pronounced the shift. Indeed, within the limits of practicable observations, the proportionality is a direct one. As one turns from one galaxy to another at twice the distance, the extent of the red-shift doubles.

Every contribution to the current cosmological argument has ultimately to be related back to this phenomenon, and the first step in the current debate is to suggest a physical cause for it. The simplest and most widely-accepted explanation puts it down to the 'Döppler effect': the increasing wave-lengths of corresponding spectrum-lines are taken as evidence that the more distant galaxies are receding from us at proportionately higher speeds. On this interpretation, all the galaxies within range of observation are scattering from one another at speeds directly proportional to their separation—like the fragments of an exploding grenade. Suppose we accept this 'recession of the nebulae' as a genuine phenomenon, what are we to make of it? Here, the rival systems of cosmology begin to diverge.

On the first account, this 'explosion' represents the complete temporal history of the entire spatial universe. In the beginning, there was a single 'primaeval atom', which encompassed within itself all the matter and energy now scattered throughout the cosmos. The creation of this 'atom' initiated the historical sequence of events through which the universe has subsequently been passing; and, for some cosmologists, its creation represented 'the beginning of time' also. Such a concentration of matter and energy was physically unstable. At the zero of time, its effective temperature was infinite, so things began to happen very quickly. After the first five minutes of history, the temperature within the exploding cosmos was a billion degrees; by the end of twenty-four hours, it was down to forty million, and all the heaviest chemical elements had been consolidated. Thereafter, the explosion proceeded for 300,000 years, by which time the average temperature had dropped to 6000°C., corresponding to the present surface-temperature of the Sun. Eventually, after 250 million years, the fragments of the initial 'grenade' were far enough apart for gravitational interaction to set in train the processes of stellar condensation and evolution explained above. This, in outline, is the 'Big Bang' cosmology as originated by the Abbé Lemaître, with elaborations by George Gamow and Edward Teller.

As against this Big-Bang view, there is the Steady-State view, whose best known advocate is Fred Hoyle. Here, the essential starting-point is matter diffused throughout the universe in a state of extreme rarefaction, and coming into existence at all times. Hoyle answers Lemaître in the same way that Hutton answered Werner: the scattering of the galaxies is only one of two counter-balancing agencies which, throughout the history of the cosmos, are for ever working against each other. At every moment in time—and there is no 'beginning' to time—the galaxies which have already condensed are scattering away from one another with increasing speeds and disappearing, as a result, beyond the reach of observation: at the same time, a corresponding quantity of primaeval hydrogen is coming into existence, distributed uniformly throughout the universe, and from this fresh star-clusters and galaxies are forming (Plate 11), so redressing the loss of those which have receded too far to be observed. Hoyle has played a leading part in unravelling the life-cycle of the stars, so the mechanisms invoked in his cosmological theories are plausible enough. At the same time, one must point out that the advocates of the Steady-State theory have gone beyond the normal limits of hypothesis, in arguing for a 'perfect cosmological principle' reminiscent of geological 'uniformitarianism' in its most extreme form. Hoyle, like Hutton, sees in the astronomical facts 'no Vestige of a Beginning—no Prospect of an End': his use of the cosmological principle implies, not merely that similar physical mechanisms have operated at every stage in cosmic history, but that, in all its leading respects and at *all times in the past*, the aspect of the heavens available to astronomical observation must have been the same. Always, the same number of galaxies have been receding at roughly the same speeds, and there have always been the same number of fresh condensations taking place.

The third view combines the pieces of this intellectual jig-saw somewhat differently. It accepts, as wholeheartedly as the Big Bang theory, that the current explosion of the galaxies represents the basic truth about the present phase of cosmic history. But it goes back beyond the moment at which the matter of the universe was at its greatest concentration, and denies that the present phase represents the totality, either of time, or of cosmic existence. The concentration of galaxies into an exploding 'grenade' some 10,000 million years ago represents only one of two extreme conditions, between which the universe is continually oscillating. The current expansion will go on until it loses all momentum, and the cosmos approaches the opposite extreme of maximum rarefaction. Once that extreme has been reached, the process will be reversed, the galaxies will begin to collapse together once more, and eventually they will re-form the intensely hot grenade. This will again be unstable and explode . . . and so *ad infinitum*.

Truth, Hypothesis or Myth?

At this point, we must ask ourselves dispassionately: How are these rival cosmological theories related to the supporting evidence—which in this case means, first and foremost, the red-shift phenomenon? This area of speculation is, of course, an alluring one. The hope that a few straightforward astronomical observations might solve perplexities about the

Creation which have exercised men's minds for so many centuries adds a natural spur to controversies which in any case have a great intrinsic interest. Yet, looking back over the whole development of historical thought, one has to recognize that the present situation in cosmology is not entirely unique. Indeed, it displays striking similarities to theoretical situations which in earlier centuries afflicted other sciences, as they became historical; and there are several signs that, in the current debate, speculation is still essentially philosophical rather than scientific. Ever since the earliest days of rational enquiry, all attempts to achieve intellectual command over the course of historical events have tended to produce theories with one of four contrasted patterns: the three *a priori* patterns associated with Plato, Aristotle and the Stoics, and the fourth, developmental pattern looking back to Anaximander and the Ionians. The most striking thing about cosmology, at the present time, is that *all* these intellectual patterns are reappearing.

The primaeval-atom theory of the Abbé Lemaître has a direct intellectual ancestry which links it to the Creation stories of early mythology, by way of Christian theology and Plato's *Timaeus*. Hoyle's Steady-State theory likewise invokes principles familiar in other areas of historical thought since the time of Aristotle: Hutton and Lyell, for example, based their arguments on the same axiom, that a rational history of Nature must presume a stable and unchanging Order. Between contemporary theories about an oscillating universe and the Stoic theory of a Cyclical Cosmos the similarities are even closer. In one respect, this last resemblance may even be a coincidence: the essential thing is not the fact that both theories involve, at one extreme, destruction of the present order by intense fire or radiant heat, but the pattern of a periodic recurrence within the temporal sequence. Details apart, the *general* resemblances between twentieth-century cosmology and its ancestors are no mere coincidence. Rather, they prompt one to look for an equally general motive. Is it, for instance, the case that, when evidence about the remote past is too slender for an empirical reconstruction of earlier history, the human intellect—for want of anything better—falls back naturally on these three *a priori* patterns of theory?

There are other signs that cosmological theory is still basically philosophical, for certain fundamental problems and objections still face us which cannot be evaded by dressing them up in twentieth-century terminology. These were clearly stated by Immanuel Kant, who realized from his own experience just how obstinate and insoluble they are. In his early years, he made Newtonian physics the basis for an evolutionary cosmology as audacious as anything proposed today; yet he later came to doubt whether the laws of physics could legitimately be extended to cover the whole of Space, Time and Matter. In particular, as he argued in his

Critique of Pure Reason, all questions about 'the Beginning of Time' or 'the Boundaries of Space' were bound to land one in paralogisms—fallacious pseudo-problems created by pressing our concepts beyond their meaningful area of use. Yet often enough contemporary cosmologists either seem to be ignorant of Kant's argument or fail to appreciate its force: they write blithely about the regions 'outside Space', or describe the Creation both as 'the beginning of time' and as having been 'preceded' by 'a timeless period of indefinite duration . . . before time was'. One does not have to be an Aristotle or a Kant to spot the internal inconsistencies in formulations such as these. Using those grand abstractions 'Space' and 'Time' as we do, one thing can be 'outside' another only if it is 'in Space'—for the phrase 'in Space' means only 'having a spatial location', not 'inside an all-embracing container'. Nor does it make sense to speak of anything 'preceding the beginning of time', for the only things which can 'precede' one another are those which are 'in time'.

If we sometimes ignore Kant's paralogisms, this may be because we forget that, in certain crucial respects, the concepts we use in mathematical physics are our own creations. Compare the notions of Space and Time with that of Temperature. We begin with an intuitive recognition that some things are warmer, others cooler; and, for scientific purposes, we introduce a numerical measure of 'temperature', which in due course acquires a particular significance and definition, within the conceptual system of our physics. So far as our everyday ideas carry us, the scale of possibilities is unlimited: however hot or cold we may make things, we can always conceive them being made yet hotter or colder. When we pass from these everyday notions to a theoretical definition of temperature, this unbounded character may or may not be preserved, and this depends on certain *intellectual decisions* which are taken—either consciously or unthinkingly—in the course of constructing our physical theories. (As a matter of historical fact, the current notion of 'absolute temperature' does not retain this unboundedness. Lord Kelvin defined it so that it had a lower limit at the so-called 'absolute zero'. Yet an alternative definition could quite legitimately have been adopted, without doing the slightest violence to any of the *facts*. A scale of 'absolute temperature' could have been so introduced that there was no lower limit, and this is in fact sometimes done, by replacing the present 'absolute temperature' by its logarithm: '0°A' thereupon becomes 'minus infinity°L'.) The general danger of which Kant was clearly aware is that of mistaking intellectual barriers which we erect—e.g. the absolute zero of temperature—for physical obstacles or limits having some *reality* in Nature.

This has an immediate relevance to the ideas of Space and Time. Much of cosmological theory today is concerned quite explicitly with 'world-models', usually based on one or other of Einstein's two theories of relativity. In constructing such world-models, theoretical physicists have to decide what theoretical definitions shall be adopted as the mathematical counterparts to our everyday notions of location and duration. Once again, these decisions can be taken so that the historical sequence of cosmic events is imagined as extending without limit in both directions—and the question 'before that, what?' always makes sense. Or alternatively, the theory may be set up in such a manner that, by definition, the temporal sequence is given an arithmetical origin or first point—then there is an 'absolute beginning of time' analogous to the 'absolute zero of temperature'. If this second choice is taken, however, we must recognize it for what it is; for we shall thereby have forfeited the right to talk about events 'before' the absolute beginning. Like those dream-temperatures below the absolute zero, these events will be turned into conceptual nonentities.

It is not always clear how far physical cosmologists see the construction of world-models as involving the extension of our intellectual concepts and how far they regard it as an exploration of the truth about Nature. Certainly, they sometimes discuss the choice between a theoretical time-scale with an absolute 'beginning', and one which is boundless, as if the existence of a 'beginning of time' were a straightforward question of fact. Whatever else it may be, it is certainly not *that*, and the introduction of theological preferences into the discussion only confuses the issue further. Both the Abbé Lemaître and Sir Edmund Whittaker frankly preferred the Big-Bang picture because it could be reconciled with religious teachings about the Creation more satisfactorily than its rivals. Many people may question whether the similarities between the 'primaeval-atom' theory and the first chapter of *Genesis* are sufficiently close to have any significance, yet the arguments which Aquinas deployed against Aristotle are evidently still influential: Pope Pius XII, for instance, welcomed the assurance that modern physics corroborated the doctrine of the Creation, even though its date had been put back from some six thousand years in the past to rather more than 6000 *million.*

Finally, the very scantiness of the astronomical evidence on which cosmological speculations are necessarily based is a further reason for caution. By itself, the evidence of the red-shift falls far short of establishing conclusively even that the nebulae are receding; so it is far from certain that, 10,000 million years ago, all the matter and energy in the heavens were concentrated in a state of extreme compression, and the further step of dentifying this moment of maximum compression as 'the beginning of time' is questionable for logical reasons. If, for purposes of argument, we accept the red-shift as evidence that the galaxies *are* receding from us, this still proves only that they are, among other things, receding from us: they are too far away for any transverse motion—across the line of sight—to be measurable and, for all we can tell at present, they may equally well be spiralling outwards from a central core, rather than scattering in straight lines from a single point. Moreover, this recession has been studied for only thirty years, and may yet prove to be a local feature of the universe operating only within the region to which our existing instruments admit us. To call the totality of the galaxies we can now study the universe-as-a-whole does not *make* it so. It draws only a verbal boundary, contingent on the capacities of our present instruments, and it would be as hazardous to guess what lies beyond these few thousand million light-years as it would have been in 1600 to guess what Galileo, ten years later, would see through his telescope.

This argument is equally applicable to the 'totality of time'. Even if both the recession of the galaxies and the phase of greatest compression are genuine, it is possible that this compression was neither absolute nor cataclysmic: it may have been merely one in a series of astronomical phases, which developed smoothly and naturally out of one another in succession—like the corresponding phases in the history of the Earth and the evolution of species. By accepting an earlier phase of high compression, one is not compelled to regard it as the scientific counterpart of the Biblical Creation. Just as we assume that the matter from which the Earth and the other planets formed had some prehistory in an earlier epoch, so likewise we have as much reason to think of the compression of the galaxies as a passing phase, forming one stage in a longer historical development. This does not mean that we must necessarily trade in the Big-Bang view for the Cyclical-Cosmos view, but rather that we must admit that our understanding has inescapable limits. We have extremely little evidence about the period from 5000 to 10,000 million years ago, and for the phases of cosmic history preceding that we have no evidence whatsoever. Once again, to call the moment of maximum compression the 'Beginning of Time' draws another verbal boundary at the limits of our present knowledge; and whether we draw this boundary 6000 years ago or 10,000 million, it is a needless hostage to fortune.

To sum up: the interpretation of the red-shift involves so many assumptions that it demands scrupulous care and unsparing logical scrutiny. For cosmology is a field in which one can very easily slide from legitimate hypotheses into 'scientific mythology', imposing an arbitrary pattern on the History of Nature, instead of patiently elucidating the temporal flux in the world around us. Where the intellectual stakes are so high, the need for caution is that much the greater, since the implicit promises that the subject may settle perennial religious doubts may prove illusory. Kant's warnings are still relevant today. As we wrote elsewhere:

> At the present stage in physical cosmology, it is no good our being in a hurry. As in history and archaeology, the only hope of solid

results lies in a step-by-step advance. There will always be some point in our map of the past history of the universe beyond which we have nothing particular to show. The same goes for the future—and how could it be otherwise? If we force astrophysics to serve us with a revised version of Genesis and Revelation we dig a pit for ourselves.

Suppose we look at the map of an unsurveyed country, we shall find that it is not completely empty, for it will bear at least the parallels of longitude and latitude. But it would be a mistake to think that these, by themselves, told us anything geographical about the surface of the earth; and the same pitfall awaits us in cosmology. However we may argue about our choice of time-scales, these will remain the bare scaffolding of astrophysical theory. The danger is that we may misinterpret them as giving us genuinely historical information.

In all other historical sciences, the crucial transition to the phase of cumulative advance has been marked by the changeover from *a priori* patterns of theory to an empirical, developmental method of enquiry. It may be that physical cosmology, with its all-embracing scope, can legitimately claim to be an exception to this general rule; but then again it may not. Just as a moment came when geologists had to leave the over-simplified theories of Hutton and Werner, abandoning both pure catastrophism and pure uniformitarianism, so perhaps the time is coming when physical cosmology too can throw away its intellectual crutches, and concentrate on multiplying our empirical knowledge of earlier phases in cosmic history. In 1784, William Herschel called on his fellow-astronomers to think of the heavens

> as a naturalist regards a rich extent of ground or chain of mountains, containing strata variously inclined and directed, as well as consisting of very different materials.

Perhaps we should now take this hint, and pursue the geological analogy one stage further. For geology finally gained an intellectual hold over the succession of past time by following up William Smith's insight into the principles of stratigraphy, and an analogous possibility is apparent today in cosmology.

The more distant the galaxies, the longer it is since their light began its journey. Accordingly, if one considers in turn galaxies at different distances, their light-spectra should reflect the state of the cosmos at correspondingly earlier phases of cosmic history. (In this way, we 'see the past' in a more direct sense than is possible in geology, biology or human history.) Hitherto, only one systematic difference has been recognized between the light-spectra from galaxies at different distances, and that is the red-shift. Yet, if the entire cosmos *is* developing from one epoch to the next, a further scrutiny of light from more and more distant galaxies should reveal evidence of this development, and on the basis of such differences we might begin to build up an empirical, stage-by-stage history such as is now taken for granted in other historical sciences. And if, however much we scrutinize our observations, no such differences appear, this itself would be significant, providing an empirical demonstration (rather than an *a priori* argument) that the astronomical cosmos has remained, for at least 1000 million years, more uniform and changeless than the objects studied by any other natural science.

It would surely do no harm now to make cosmology more empirical, working back from our knowledge of the evolution of stars and galaxies to valid conclusions about cosmic history, instead of jumping straight to the Creation. We think of the origin of society, Man, life, the Earth, the Solar System, star-clusters and even galaxies as involving gradual and continuous transformations taking place over a period of time: our ideas about them are no longer cataclysmic or *a priori*. No doubt, current theories in physical cosmology have the same attraction that extreme uniformitarianism had for Charles Lyell in his early years—so that, having accepted Darwinism, he could say, 'It cost me a struggle to give up my old creed.' It has always been a struggle for men to renounce the apparent short-cuts which *a priori* patterns of theory offer them, and to build up a genuinely empirical account of historical change and development laboriously and without preconceptions. The question today is whether the persistence in physical cosmology of the old familiar intellectual patterns can be justified by the unique character of its subject-matter; or whether the same struggle will soon be demanded of cosmologists also.

Are the Laws of Nature Changing?

One further point must be mentioned in conclusion. We saw earlier in this chapter how the permanent framework of Newton's physics, with its five distinct categories of changeless entity, was progressively eroded, so that by the mid-twentieth century only the 'laws of nature' defining the fundamental 'force-fields' still preserve their timeless character. But, looking at these present-day heirs of Newton's laws of motion and gravitation, one must ask one further question: Will these laws of nature remain for ever as the unchanging framework of the physical scientists' world-picture? This question arises both as a natural extension of the historical changes we have been studying, and because it has a direct connection with the present phase in cosmological thought.

To relate it back to our earlier discussion: the phrase 'laws of nature' acquired a central position in European natural philosophy only in the seventeenth century. It entered the European tradition in a very particular historical context, and it had from the outset certain specific theological associations. The men who conceived the mathematical formulae concerned called them 'laws' because they saw in them the 'rules' by which the 'behaviour' of material things was 'governed'. At this initial stage, the legal analogy implicit in the word 'laws' was something more than a metaphor; and, since God Himself was regarded as changeless and eternal, it was presumed that the 'laws of nature' which were an expression of His will were correspondingly fixed in their form. So the rational framework of the natural world which they called 'the Order of Nature' was defined, and in justification it was sufficient for Newton that

> It became Him [God] who created all material Things to set them in order. And if He did so, it's unphilosophical to seek for any other Origin of the World.

The Order of Nature was frozen by God's rational choice.

By now, however, this particular theological framework has lost its hold over us. Like so much else in the static world-picture of the seventeenth and eighteenth centuries, the eternal fixity of physical laws can scarcely be assumed any longer without examination. In any case, there are some scientists who have positive reasons for believing that this fixity, too, may be an illusion. Certainly, if the laws of nature were changing slowly enough, there would be no practical way of detecting this fact—even over millions of years—and in that case the question whether they were changing would be purely academic. But perhaps these changes in the physical processes occurring at different cosmic epochs are, after all, *not* undetectable; and perhaps we may even have already in our grasp—unrecognized and unappreciated—the evidence to show their nature.

The red-shift in stellar spectra might itself be just that evidence, and is interpreted as such in one of the more unorthodox explanations offered for it. The more distant the galaxy from which a light beam comes, the more remote in time were the physical processes by which it was emitted. Now, if the numerical constants in the laws of nature governing these processes 'drifted' slowly with the passage of time—e.g. if Planck's constant, h, were slowly increasing—this would produce the familiar red-shift, even though the distant galaxies were not receding at all. The apparent 'recession of the galaxies' would be an illusion, produced by slow secular change in the laws of nature, and the 'primaeval atom' would disappear from cosmological history.

This possibility is mentioned, not because the positive evidence in its favour is particularly strong, but simply to complete a balanced account of ideas about cosmic history. At a theoretical level, a Universe whose laws are changing from epoch to epoch is no longer an idle fantasy. The physicist P. A. M. Dirac, for example, has been exploring the intellectual consequences which follow when physical forces of different kinds (e.g. electrical and gravitational) vary slowly in their relative strengths: in this way, he links the atoms, with their electrical forces, to the galaxies, with their gravitational forces—the microcosm to the macrocosm—in a manner which changes through cosmic history.

We saw in *The Architecture of Matter* that the future development of sub-atomic physics is as open today as it has ever been. Nor can we be any more certain about the pattern of future cosmological theory. Though, during the last fifty years, our understanding of stellar evolution has been a great intellectual advance, which in some form or other will certainly survive in our world-picture, many profound and difficult questions are still not even half answered, and one of these is central: Do all the fundamental laws and constants of physical theory retain their forms and values eternally, without any modification? About this we can no longer be certain. We may yet be on the threshold of the greatest of all scientific revolutions.

FURTHER READING AND REFERENCES

The best popular introduction to contemporary theories of cosmology and stellar evolution—including Dirac's 'variable laws'—is

J. Singh: *Great Ideas and Theories of Modern Cosmology*

The astronomical background is presented in a number of recent books: for example

W. Baade: *Evolution of Stars and Galaxies*
O. Struve and V. Zebergs: *Astronomy of the 20th Century*

There is a sound new biography of William Herschel by M. A. Hoskin.
Most of the participants in the cosmological debate have published general books on the subject—notably

H. Bondi: *Cosmology*
G. C. McVittie: *Fact and Theory in Cosmology*
G. J. Whitrow: *The Structure and Evolution of the Universe*

See also Whitrow's book *The Natural Philosophy of Time*, and the admirable critique of cosmology in the book

M. K. Munitz: *Space, Time and Creation*

The most useful collection of original papers in this field from the earliest times is the book

Theories of the Universe: ed. M. K. Munitz.

EPILOGUE

Nature and History

IN THE second half of the twentieth century, we find ourselves caught up in an intellectual transformation which is still incomplete. Since 1800, the discovery of unlimited time has fractured the foundations of earlier world-views even more irreparably than the earlier discovery of unbounded space. Bit by bit, the static framework of the natural world, on which Protestant natural theology rested, has been dismantled; and by now, the 'ineffaceable characters' of atoms and molecules, which Maxwell saw as the ultimate justification of 'accuracy in measurement, truth in statement, and justice in action', have dissolved into a temporal flux billions of years in length. Permanence—as Plato hinted all along—belongs to the conceptual world alone.

This discovery of time has had four aspects. To begin with, it has compelled men to recognize the sheer *extent* of the past—something which they might guess, but before the eighteenth century could never prove. Secondly, we can now establish the true antiquity of the world, just because we have also come to understand the general processes by which it has developed: the time-scale of these *developmental processes* provides our final measure of the past. Thirdly, these processes have, at every level, been of two kinds—some general and repetitive, others cumulative and progressive—and a central problem for any historical enquiry, whether about Man or Nature, is to distinguish *universal* factors, operative everywhere and at all times, from *conditional* ones, which arise out of earlier historical phases and so acquire significance only at a particular stage of temporal development. Finally, there is the question of *motivation*: in human history, at any rate, full understanding comes when we reconstruct for ourselves the reasons, fears and ambitions of our fellow-humans caught up in historical events.

With these four aspects in mind, we can now resolve a question raised at the very beginning of this book: Can one legitimately speak of the development of Nature as 'history' at all? In three of our four respects, the ideas and methods for interpreting temporal development apply equally to human history and to natural history. The antiquity of Humanity, like that of Nature, has been carried back far beyond Adam and Eve. Human customs and institutions, like organic species and stellar galaxies, have proved to be temporary products of continuing processes of development. The contrast between the recurrent and cumulative elements in time creates its problems for the human historian and the natural scientist alike: every change in time builds to some extent on novel features of the past, and creates fresh elements of novelty for the future to build on in its turn. To that extent, all temporal development embraces a sequence of phases, each of which creates a novel starting-point for its successor, and 'progression'—as distinguished from 'progress'—is inevitable.

In one respect alone can a sharp distinction be drawn between the course of human affairs—'History', in its narrow sense, and the temporal development of Nature—traditionally known as 'Natural History'. This distinction arises over the question of *motivation*. Even the story of men and their societies cannot be entirely explained in terms of human motives alone: the evolution of Rome (to adapt one of Collingwood's own examples) involved many other factors besides the plans and passions of the Romans and their rivals—the changing fertility of the Campagna, climatic variations in North Africa, and the growing population of malarial mosquitoes in the Pontine marshes. Still, as historically-minded philosophers have emphasized from Vico on, we can achieve a sympathetic understanding with the men of earlier times which we cannot aspire to do with mosquitoes or rock-strata. Where differences of culture and custom are great, this may be very difficult; but, when we succeed in seeing our forefathers' choices and decisions as they did themselves, we undoubtedly gain an 'inside' view of history quite unlike anything open to us in geology, cosmology or evolutionary biology.

Even here, perhaps, the distinction is no longer absolute. As ethology improves our understanding of animal behaviour, we may acquire something of the same insight into the outraged feelings of territorial possession that set the Roman geese cackling on the Capitoline Hill. Human motives, after all, are not entirely unique. Thus, in the last resort, what marks human history off from natural history is the fact that it is—quite deliberately—man-centred. The motives of the Capitoline geese are interesting to the Roman historian only for their effect on the human city: we can leave the geese themselves to write their own 'goose-history'.

Taking Nature and Humanity together, we now recognize that, in almost every respect, the present state of affairs has developed naturally out of a very different past. It is true that there are still gaps in our account. If we go back far enough, we reach a screen of ignorance on to which we can do no more than project *a priori* forms of theory shaped in our own minds. We can speculate a little about the possible origin of life on Earth: about the evolution of human society from pre-human gregariousness we are hardly even ready to speculate. Nevertheless, what has already been achieved was far harder than what remains to be done, and there is reason to think that the worst gaps in our knowledge will in due course be filled by methods that have already paid dividends elsewhere. The origin of life, for instance, is no longer thought of as a matter for *a priori* argumentation. We have moved from the question: 'Has life always existed in some form or other, or was it created at some definite moment in the past?', and now ask, rather: 'By what slow sequence of changes in the pre-biotic environment did the essential constituents for single-celled organisms come into existence and survive?' This reformulation is not idle, for men such as Oparin, Urey and Calvin are already on the way towards answering it. Ultra-violet light is apparently capable of forming complex chemical compounds out of suitable gas-mixtures, and Calvin has even argued in terms of a primitive 'natural selection', which favoured the original accumulation of chemical raw materials for later cells and organisms.

Treating the whole temporal development of the world as a field for historical enquiry can yield a further dividend. As one field after another has been opened up, there has been a striking continuity of problems and methods. Where questions of motivation are not crucially involved, the historical inferences by which we reconstruct the past show a common form both in different centuries and in different historical sciences. Consider the role of experiment in history. There is, of course, no question of experimenting directly on the past. Nonetheless, experiments and observations made at the present time have frequently shown how things *may have* happened—even, how they *presumably did* happen. Calvin and his collaborators today are working in a tradition established by Buffon in the eighteenth century, and their experiments on the effects of radiation on gas-mixtures have a similar relevance to the origin of life that Buffon's experiments with cooling metallic spheres had to the origin of the Earth. Studies of this kind provide the same circumstantial evidence for science that ballistic experiments with a murderer's bullet provide for criminal investigations. Establishing how a crime could have been committed is an important step towards deciding how it in fact was committed. Likewise, in human history: observations and experiments carried out today may have the same indirect bearing on the historic past as they do on the prehistoric epochs of geology and astronomy. Were the communal frenzies reported in the Middle Ages caused by ergotism—that is, by eating bread made from a particular kind of diseased grain? How do mass-unemployment and near-starvation tend to affect men's political attitudes? Did Napoleon die of arsenical poisoning? In each case, the better we can match our present knowledge to the evidence of the past, the more we can improve our understanding of earlier human history.

Historical understanding comes through exploiting continuities between the past and the present—matching the current patterns of relationships against what we know of earlier events. So the idea of historical *dis*continuities is as unenlightening in the human realm as it is in natural science. The concept of 'revolutions' has the same defects for the purposes of political history that the concept of 'catastrophes' has always had for natural history. Only when we learn to relate the uniformities underlying political or geological change—however rapid—to processes or activities familiar to us in our own time, do we get a grip on the past.

The Evolution of Ideas

Whether we are concerned (as in *The Fabric of the Heavens* and *The Architecture of Matter*) with the overall structure and layout of the universe, or the different kinds of material body and their interactions, or (as in this book) with the character of the historical process, we find the same recurrence of intellectual patterns. In each case, we have concentrated on a single intellectual genealogy, and tried to show how our present-day ideas are related back by an intelligible ancestry to earlier conceptions. If certain family likenesses have persisted throughout each genealogy, this need be no surprise. Given the continuing similarity of the problems with which men have been concerned, a corresponding recurrence in their forms of theory should perhaps be expected. For scientific thought has advanced, not by men's stripping veils from Nature and reporting on what they see, but through a continuing effort to construct more consistent, precise and adequate conceptions of an invisible reality. As the Count in *Prospero's Cell* (Lawrence Durrell) insists, this does not mean regarding 'the hard and fast structure of the sciences' as providing 'nothing more than a set of comparative myths, some with and some without charm': rather it is to recognize that 'there is a morphology of forms in which our conceptual apparatus works'. The scientific ideas of a given age are much more than a bald portrait of Nature: they reflect men's untiring desire to match their forms of thought against the testimony of things.

> 'All this is metaphysics,' says Zarian a trifle unhappily.
>
> 'All speculation that goes at all deep,' replies the Count, 'becomes metaphysics by its very nature; we knock up against the invisible wall which bounds the prison of our knowledge. It is only when a man has been round that wall on his hands and knees, when he is certain that there is no way out, that he is driven upon himself for a solution.'

The old quandaries at the heart of Greek metaphysics, about the relation of concepts to objects, also yield up some of their mystery if reconsidered in the light of intellectual evolution. In our three case-studies, we have followed out the development of ideas both in sciences which create idealized mathematical concepts, and in other historical sciences, where these abstract conceptions come face-to-face with the actual sequence of events. As a result, we have been forced to recognize just how far the concepts and axioms of physical theory are indeed 'abstract'—intellectual patterns sifted from the flux of Nature, and selected as 'paradigms'—standard examples by which more complex phenomena are explained in their turn. Still, as the physicist Werner Heisenberg has clearly recognized, this process of 'abstracting' creates a logical gulf between physical theory and Nature, between 'natural philosophy' and 'natural history':

> The discovery that the earth is not the world, but only a small and discrete part of the world, has enabled us to relegate to its proper position the illusory concept of an 'end of the world', and instead to map the whole surface of the earth accurately. Likewise modern physics has purged classical physics of an arbitrary belief in its own unlimited applicability. It has shown that some parts of our science—e.g. mechanics, electromagnetism and quantum theory—present scientific systems complete in themselves, rational, and open to exhaustive investigation. They state their respective laws of nature, probably correctly, for all time.
>
> The essence of this statement lies in the phrase 'complete in themselves' . . . Within a system of laws which are based on certain fundamental ideas only certain quite definite ways of asking questions make sense: thus, such a system is separated from others which allow different questions to be put. So the transition in science from fields of experience already investigated to fresh ones never consists merely in applying laws already known to these new fields. On the contrary, a really new field of experience always leads to the crystallization of a new system of scientific concepts and laws.

The self-sufficiency of a mathematical theory such as Newton's dynamics or Maxwell's electromagnetism is purchased at the price of a certain detachment from the world of fact. Once we have framed such a theory, we are always justified in asking just how widely this particular intellectual pattern is relevant, and to what kinds of phenomena. This contrast between the neat, self-contained World of Ideas and the flux of the natural world explains why single-minded physicists find natural history an intrinsically messy subject. Meteorology is bad enough: there, the very multiplicity of factors frustrates the theorists' aesthetic preference for elegant and rounded solutions. But in every kind of history one must face, in addition, the eternal unpredictability of novelty. Science as a whole can never be 'complete in itself', in the way that a single branch of physical theory can. To that extent, Nature always retains the power to spring surprises. So the scientific priority and superiority sometimes claimed for physics—as laying the intellectual foundations, first for chemistry, and subsequently for physiology and even psychology—depends on our taking up an abstract, Platonizing point of view. When, by contrast, our search for understanding forces us to look at Nature in a more historical, Aristotelian light, the 'peck-order' is reversed: the science of palaeontology first achieved maturity, followed by geology, zoology and chemistry, and physics is still in its infancy.

With Heisenberg's concession, theoretical physics is at last coming to terms with the historical insights of Vico. As Vico insisted, the absolute certainty characteristic of mathematical systems springs from the fact that they are our own creations—'We know them fully, because we ourselves have made them'. This is the complement of Heisenberg's observation. The World of Ideas is self-contained, cogent and certain, just because we fashion it deliberately so that our minds can move freely and confidently within it. But no system of concepts, however cogent and compact, can *guarantee* its own relevance to the flux of the natural world—that is, to the actual course of historical change. To that extent, the historical sciences are more profound and difficult than the theoretical ones, for in them we are compelled always to take the world as we find it, and cannot limit the scope of our ideas to suit our own convenience.

Science began in Ionia with the search for permanent elements behind the flux of events. From the beginning, philosophers like Herakleitos declared that no unchanging elements were to be found, outside the creations of our own minds—that everything in the natural world was in flux, while certainty and eternity belonged to the world of the intellect. Subsequent scientists have found this a hard saying: they have continued to hope for permanence within Nature, and have even demanded the presence of a fixed framework as a necessary mark of rationality. Yet by now it is beginning to appear that Herakleitos may have been right. In place of a natural order having a permanent and clearly defined structure, we are learning to accept an evolutionary process, in which all things are subject to temporal development. And if the 'laws of nature' themselves turn out to change slowly from one cosmic epoch to another, that will be only the last step in a historical transformation which has already affected the rest of science.

Fortunately earlier fears, that only a fixed Order of Nature could be intelligible, have proved groundless. In evolutionary terms, we understand the world not less, but more completely. Though the development of scientific ideas is, like everything else, a process of continuing change which can never be absolutely complete, we can draw intellectual strength from this very fact.

Bats are now beginning their short strutting flights against the sky. In the east the colour is washing out of the world, leaving room for the great copper-coloured moon which will rise soon over Epirus. It is the magic hour between two unrealized states of being—the day-world expiring in its last hot tones of amber and lemon, and the night-world gathering with its ink-blue shadows and silver moon-light.

'Watch for her,' says the Count, 'behind that mountain there.' The air tastes faintly of damp. 'She will be rising in a few moments.'

'I am thinking,' says Zarian, 'how nothing is ever solved finally. In every age, from every angle, we are facing the same set of natural phenomena, moonlight, death, religion, laughter, fear. We make idolatrous attempts to enclose them in a conceptual frame. And all the time they change under our very noses.'

'To admit that,' says the Count oracularly, 'is to admit happiness—or peace of mind, if you like. Never to imagine that any of these generalizations we make about gods or men is valid, but to cherish them because they carry in them the fallibility of our own minds.'

Index

D

M

N

O

P

Q

R

S